Elixir

Elixir

A Parisian Perfume House and the Quest for the Secret of Life

Theresa Levitt

Harvard University Press

Cambridge, Massachusetts 2023

Published in the United Kingdom as *Elixir: A Story of Perfume, Science and the Search for the Secret of Life* by Basic Books, an imprint of Hachette Book Group, London

Printed in the United States of America

First printing

Library of Congress Cataloging-in-Publication Data

Names: Levitt, Theresa, author.
Title: Elixir : A Parisian perfume house and the quest for the secret of life / Theresa Levitt.
Description: Cambridge, Massachusetts : Harvard University Press, 2023. | Includes bibliographical references and index.
Identifiers: LCCN 2022031709 | ISBN 9780674250895 (cloth)
Subjects: LCSH: Laugier, Edouard. | Laurent, Auguste, 1807–1853. | Perfumes—France—History. | Chemistry—France—History. | Asymmetry (Chemistry)—History. | Elixir of life.
Classification: LCC TP983 .L475 2023 | DDC 668/.540944—dc23/eng/20220808
LC record available at https://lccn.loc.gov/2022031709

Contents

Prologue

A New Philosopher's Stone

PARIS, RUE BOURG-L'ABBÉ, 1835

Édouard Laugier had distilled the essence of bitter almonds many times before. Although the seeds were bitter and poisonous to eat, the fragrance was soft and pleasant, making it a staple at his family's perfume house of Laugier Père et Fils. Édouard lived with his family above the shop in the center of Paris, their bedrooms and offices taking up the entire second floor. Above them on the third floor were rooms with massive cauldrons and cooling pans for making soap. Below them, on the ground floor, their shop spread over the length of two facades, its floors and mezzanine outfitted with mahogany counters stacked with ceramic pots and glass-fronted armoires displaying bottles and vials. A hallway led to another set of rooms, lit by windows to an inner courtyard, and off-limits to the public: the kitchen, dining room, and a laboratory for preparing perfume materials. It was in this last room that Édouard found himself now, on a late summer's day in 1835, readying the pressed almond cakes for distillation. This batch, however, was not destined for the store's shelves but for his own investigations into the chemistry of life.

He had not always wanted to follow the family business. He had left home at nineteen, crossing the Seine to the Left Bank of Paris, where he set up a laboratory in the shadow of the Sorbonne and tried to join the ranks of the academic chemists. It proved a tough scene to break into. The positions were all controlled by the followers of Antoine Lavoisier, whose new chemistry, with its precise chemical formulas, replaced an antiquated language of spirit and phlegm. Édouard never managed to get a steady academic position, and after a few years of barely scraping by, he returned home to Laugier Père et Fils. He brought with him a friend he had met on the Left Bank, Auguste Laurent, who had found himself in a similar situation.

And so the two began working in a laboratory in the back of the perfume shop. By day they distilled essences and mixed perfumes; at night they dreamed of solving the mysteries of chemistry. The biggest issue of the day was the ongoing effort to extend Lavoisier's chemical revolution into the domain of living things. He and his disciples had achieved startling success in the inorganic realm, neatly arranging its materials into precise formulas and balancing the ledger books of their chemical reactions. The organic world, by contrast, had remained stubbornly ungovernable, its thousands of substances appearing to be nearly indistinguishable piles of carbon, hydrogen, and oxygen.

Two German chemists, Justus Liebig and Friedrich Wöhler, had recently found a clue in the chaos. They had been trying a scattershot approach, reacting the essential oil of bitter almonds with just about everything they could get their hands on. Working backwards, sifting through their results, they identified a single configuration of atoms that seemed to remain constant through all these reactions. They named it the benzoyl radical: the first point of constancy in what had previously been a churning sea of indistinguishable carbons and hydrogens. Hailed as a point of light "in the dark province of organic nature," it was a key that promised to unlock chemistry's thorniest difficulties.[1] The problem was, no one had ever isolated it, despite the efforts of the most renowned chemists of Europe.

Édouard had not found it either, but this time he tried something a little different. As he went to soften the almond cakes before distilling them, he switched his usual water, drawn from the Seine, with water from one of the new artesian wells that had recently been drilled in Paris. A subtle change, no doubt, but these things mattered. When he made his house's signature Eau Régénératrice, which promised rejuvenating effects drawn from the "virtues of plants," Édouard always specified that he distilled the crushed bergamot peel with water drawn from the river and the bitter oranges with water from the fountain (and he continued to specify as he added Portugal oranges, mint, tarragon, cinnamon, and roses).[2] Other recipes were even more particular, insisting on water taken from the turbulent eddies beneath the blades of a mill wheel.

Switching the water paid off. There, amid the purified distillate in the distilling flask, was something he had never seen before. It was what perfumers called a "resinoid"—a viscous blob resembling tree sap. He called over Auguste Laurent, who ran it through some tests. Auguste, like Édouard, had taken a nontraditional path to the study of chemistry. His parents had originally sent him to mining school, hoping to see him established in a lucrative profession. While the "lucrative" part never panned out, he did learn some useful things few other chemists knew. Confronted with his mystery resinoid, he tried running a stream of chlorine over it, a procedure he had learned in his previous work with coal tar. He then dissolved the resultant product in alcohol, crystallized it, and measured the crystalline angles: a mineralogist's technique hardly used among chemists.

It was, to all appearances, the elusive radical. While the substance itself was rather bland ("light yellow, perhaps colorless, odorless, tasteless" were the notes), the crystals formed "beautiful prisms." Auguste put some in his mouth to test its properties. "When chewed," he reported, "they produce a disagreeable sensation."[3] But he could still savor the taste of victory. The shards crunching between his teeth were, for him and the rest of the chemistry world, the best hope for unlocking the secret of life.

This secret had tantalized the best minds of Europe for millennia. What separated living and nonliving matter? What made the dynamic, organized world of living things so different from the inert mineral world? This question had led Plato and Aristotle to speak of vegetable souls. It drove the alchemists' quest to distill the living spirit of plants and divide it into two groups: *spiritus vini*, or the spirit of wine, and *spiritus rector*, the "guiding spirit" responsible for fragrance. Over the eighteenth century, naturalists came to associate this guiding spirit with a vital force that directed plants' growth, endowing them with their complex organized structure.

But by the 1830s, chemists had largely abandoned the search. A new dogma had crystallized in the modern age, one insisting on an equality between living and nonliving things. They were composed of the same matter, the claim went, and followed the same chemical laws. Lavoisier had led the charge, banishing anything that hinted at alchemy or a distinct vital force. His efforts had included a complete overhaul of the language chemists used: oil of vitriol became sulfuric acid and crystals of the moon became silver nitrate, amid hundreds of other changes. When he got to the *spiritus rector*, he proclaimed with his coauthors, "we did not think we could let it survive."[4] He offered the word "aroma" as a replacement but warned that it corresponded to nothing real.

The benzoyl radical was supposed to bring the complicated world of living things more in line with Lavoisier's system, extending its compositional paradigm into the organic realm. But nature proved elusive. As Laurent and Laugier worked through the implications of their discovery, they found that it was not the definitive answer they had hoped for but merely the beginning of a series of deeper, stranger questions that cast them even farther outside the realm of mainstream chemistry. It was as if the *spiritus rector* refused to die, and instead continued to guide the organization of living things in ways that chemists could not replicate.

The pursuit of this mystery revealed a deep, unbridgeable divide between the products of the natural world and the artificial creations of the chemists, one that still stands today and constitutes one of science's great unanswered questions. Its path ran through not only the essential oil of bitter almonds but a sprawling cache of nature's most fragrant materials: the crisp scent of lavender, the soft redolence of vanilla, the sharp waft of camphor, the fresh blast of wintergreen, even the acrid stench of opium. At the center of it was a perfume house, the oldest house in Paris by the time Édouard and Auguste worked there. Édouard's grandfather Blaise had founded it over fifty years ago, selling the distilled vitality of various flowers, herbs, roots, seeds, gums, and resins to a clientele clamoring for their life-sustaining qualities. Here is where the story begins.

1

The Store of Provence in Paris

PARIS, RUE BOURG-L'ABBÉ, 1770

Édouard's grandfather, Blaise Laugier, left southern France as a young man to try his luck selling perfume in Paris, along with his wife, Marie-Jeanne, who bore their first child, a son, soon after arriving. Their hometown, Grasse, had a well-known reputation as the center of perfumery in Europe, but Laugier had been unable to crack the deeply entrenched guilds there. Paris, his new home, had a looser guild system, a clientele with ample money, and a stench of legendary renown that needed all the perfumers' arts to combat it. Its inhabitants complained of a smell "no foreign nose could abide." And the best minds of the age agreed. Its "dirty and stinking streets" were the first thing Rousseau noticed when he arrived, and Voltaire lamented its "shadow and stench" while in exile.[1]

A visit to Laugier's shop on the rue Bourg-l'Abbé involved running the gauntlet of the most distinctive odors Paris had to offer. Coming from the Left Bank, you crossed over the Seine at the Pont au Change to arrive at what Paris's most exacting chronicler, Louis-Sebastien Mercier, called "by far the worst smelling place in the world": the rue du Pied-de-Boeuf. Packed into a small square were a crowded prison, a storehouse for keeping dead corpses, a butcher, a slaughterhouse, and a filthy fish market. An open channel of human effluvia met up with a stream of blood from the slaughterhouse, where they combined to flow into the Seine, the primary source of drinking water for Paris.

FIGURE 1. The Holy Innocents Cemetery served as a site of mass graves since the Middle Ages. An ossuary for storing bones ran along the south side.

Stepping over the stream, your path continued up the rue Saint-Denis, past the Holy Innocents Cemetery, whose fetid air was already causing an outcry by the 1770s. Paris had buried the destitute here since the twelfth century, often in mass graves and covered only with shrouds. By the eighteenth century, shifting foundations had begun exposing the half-decomposed contents, with neighbors complaining of body parts breaking through their cellar walls. The stench of death permeated everything, and

FIGURE 2. The rue Bourg-l'Abbé ran between the major thoroughfares of rue Saint-Martin and rue Saint-Denis, depicted here on the Turgot map of 1739. (The streets marked Grand and Petit Heuleu on this map later went by the name "Hurleur").

the air of certain cellars was so thickly mephitic you could suffocate on entering them.

Next came Les Halles, an open food market where each stall bore its own olfactory signature. Walking north on the rue Saint-Denis brought you first past the rue aux Fers, specializing in the sale of hay, then the rue de la Cossonerie, specializing in poultry. The sellers of cheese and fish were not far off, unmistakable even at considerable distance. The sensorium of Paris changed with each corner turned. The Russian poet Nikolay Karamzin noted as much when visiting in the 1780s. In one moment, he said, "filth is everywhere and even blood is streaming from the butchers' stalls. You must hold your nose and close your eyes." But you only had to take one step farther, "and suddenly the fragrance of happy Arabia or at least Provence's flowering meadows, is wafted to you, for you have come to one of the many shops where perfume and pomades are sold."[2]

If you turned away from the open markets onto the rue aux Ours, you would soon come to the opening of the rue Bourg-l'Abbé, lined on both sides with shops catering to a rising bourgeois clientele. The street ran parallel and in between the rue Saint-Denis and the rue Saint-Martin, the two main north-south thoroughfares of central Paris. These two streets had distinct reputations: Saint-Denis was known (then as now) for its prostitutes and nightlife, and Saint-Martin for its churches and decorum. Bourg-l'Abbé, tucked between them, bridged this range of human activity. Its shops offered the respectable middle class all the wares needed for a well-organized household while also catering to the deepest impulses of luxury and desire.

Laugier's shop at number 30 was about two-thirds of the way up the street. Nestled between a florist and a shop selling scented fans, it marked a rare respite from the crushing stench of city life. Its windows looked across to the dark, narrow passageways of rue du Grand-Hurleur, notorious as a place where the world's oldest profession was plied. Yet even where the light was good, the profusion of shops gave the impression that anything could be bought with enough money. Merchandise both fashionable and fanciful filled the shops that lined the street, much of it the products of highly specialized crafts created in line with strict guild statutes. There were individual shops devoted to ribbons, paper, bonnets, jewelry, lace, instrument strings, playing cards, and more. There was one shop for suspenders and one across the street for belts.[3]

If you continued north along the street, you would reach the abbey of Saint-Martin-des-Champs, which gave the street its name. Both the abbey and the street date to at least the Carolingian period, when they were outside the stone walls that once encircled Paris. The abbey built its own protective walls, and the street found itself within the walled enclosure, or *bourg*, of the abbey. Although an expansion of the Paris city walls in the twelfth century came to include it, the name remained, its medieval roots belying the fact that the area had become a thriving, bustling commercial center.

It was here that Blaise Laugier set up shop, advertising as "the store of Provence and of Montpellier."[4] Other perfumers from the south of France would follow his path to Paris in the following decades, including such storied names as Jean-François Houbigant and Jean-Louis Fargeon.[5] But while they both chose fashionable addresses near the Tuileries and catered to the royal court, Laugier reigned over the middle-class market of central Paris, bringing to the shadowed corners of its narrow streets the sun-soaked floral bounty of his former hometown, Grasse.

THE FRAGRANT HILLSIDES OF GRASSE

The town of Grasse in Provence seemed preternaturally destined for growing flowers, nestled in an ideal spot between the foothills of the Alps and the Mediterranean Sea. Its long, narrow terraced slopes face southeast to receive as much sun as possible while remaining protected from the dry mistral winds coming off the water. The unusually even climate guarantees a consistency from year to year that allows a wide range of different flowers to bloom in successive waves. But these endless fields of flowers, which still cover the hillsides today, were a recent addition in the eighteenth century, brought in to cover up far less pleasant smells.[6]

Grasse's original reputation was for the most repulsive-smelling of industries: leather tanning, a process that involved, at various stages, soaking the skins in stale urine, pounding them with dung, and allowing an extended period of "bating," or supervised rotting. The combination of excrement and decaying animal flesh was so overwhelming that most towns relegated their tanneries to the far outskirts. But Grasse, whose numerous springs provided the necessary supply of water, made tanning its central enterprise. Its reputation for making leather stretched back to the Middle Ages, and by the end of the sixteenth century, Grasse was known for its fine leather gloves.

It was also in the sixteenth century that a few Grasse tanners began treating their gloves with flower petals to mitigate the lingering smell. The practice had begun in Italy and had come to the French court with Catherine de Medici when she wed the French king. At her insistence, glovers in Grasse set up laboratories to duplicate the expensive perfumes being imported from Arabia and Spain.[7] At first, they drew from the local flowers, primarily lavender and aspic (sometimes called lavender spike). But the countryside was soon transformed. The French East India

Company, founded in 1664, brought back fragrant plants from around the world, including jasmine from India and the "perfume rose," a rose smaller but more fragrant than the native varieties.[8] The 1670s saw the planting of tuberose, a sweet, roselike flower from Mexico. Monks from the nearby abbey of Lérins brought in the bigarade, or bitter orange tree, whose flowers produced the highly desirable neroli oil.[9]

By the eighteenth century, perfuming had eclipsed tanning. The guilds separated in 1724 when Jean Galimard, a tanner who had become the purveyor of pomades and perfumes to King Louis XV, organized a separate guild of *gantiers-parfumeurs*, or glovemaker-perfumers.[10] They established a coat of arms of their corporation—a glove and two circles—and a strict set of statutes that made it very difficult to join. Only twenty-one men were permitted the title of "master glovemaker-perfumer," with no new applicants allowed. A change in the leather laws made tanning impractical in the city, further shifting the balance from gloving to perfuming.[11]

With the urine pits closed, the air around Grasse took on a sweeter aspect. Villagers claimed they could pinpoint the time of year by the smell, as wave after wave of flowers blossomed in the fields. The first to bloom, in February, were the mimosas, a fleeting explosion of yellow across the hillsides whose soft, honeyed, powdery scent marked the end of winter. Next were violets in March, jonquils in April, orange blossoms and roses in May, and tuberose starting in June. Jasmine was the season's show-stopping finale, blooming from August to October, and requiring some of the most intensive and carefully orchestrated collection. Each flower had its favored moment to exude its essence, and for jasmine it was the first moments of dawn. Roses, by contrast, were most fragrant in the late afternoon, and could be picked only then. Geranium, mint, lavender, iris, hyacinth, sage, cassia—each required its own approach.[12]

By the eighteenth century, the operations had grown to an unprecedented size and complexity, making Grasse the first place to manufacture perfume on an industrial scale. Gone were the days of simply laying flower petals on top of leathers. They now had a host of different procedures in place to pry the fragrant oils from the plant. No one procedure worked in every case, and with dozens of different plant species, Grasse perfumers employed a variety of techniques, some of them stretching back millennia and others newly developed for the purpose.[13]

The oldest process, known as expression, was to physically squeeze the oils out. This was not complicated and had been practiced for most of human history, but it generally worked for only the most robust materials, such as citrus peels and seeds. In Grasse, this included the important perfume ingredients of orange, lemon, grapefruit, and bergamot (the latter being an essential ingredient of Catherine de Medici's favorite *aqua de regina*).

Another method of only limited use in Grasse was steam distillation. Steam could separate the volatile oils from the rest of the plant material, but few plants held up well under the process, and these were usually the hardier herbs that often carried scent in their stalks, such as rosemary, thyme, and peppermint, as well as cloves and juniper. Of the much-desired florals, only lavender distilled well with steam. The process was relatively straightforward. After being cut, the plants were left to wilt for a bit, and packed onto a rack located above a boiler. As the water boiled, the steam rose up through the plants, taking with it the volatile oils on the surface of the plants. The steam then cooled in a condenser, and the water and oil separated into two immiscible distillates. After waiting for them to separate, one could pour the oil off the top.

More delicate plants required the process of enfleurage, which coaxed a flower's scent into a fat, where it could be preserved. The basic principle had been in practice for millennia, with unguents and scented oils comprising an important trade in the Mediterranean from at least the time of Nefertiti. But the perfumers of Grasse scaled up these techniques, creating an enterprise of vast human labor that transformed the way perfume was made.

The most common technique was "enfleurage à chaud," sometimes called digestion or maceration. A purified fat (usually beef or deer suet) or oil was melted in a bain-marie, a water-bath that allowed for slow, controlled heating below the boiling point of water. The perfumer would

FIGURE 3. A scene from Grasse depicting maceration, or "enfleurage à chaud," in which flower petals are heated in a vat of oil.

put the flowers in the liquefied fat, leaving them there at a gentle heat until they were spent of odor (usually between twelve and forty-eight hours). Then they drained the spent flowers from the fat and added new ones, repeating the process ten to fifteen times. The end result was a pomade, graded with a number that was supposed to represent the ratio of the weight of the flowers used in production to the weight of the fat.

But the technique that established Grasse's reputation was "enfleurage à froid," an even more painstaking process that achieved industrial scale only in Grasse. It was reserved for the flowers whose scent could bear no heating at all, which happened to include two of the perfumers' great superstars: jasmine and tuberose. Jasmine pickers would head to the field in the moments before dawn, when the flower was most fragrant. They had to be careful not to damage the petals in any way, as that altered the scent. After the baskets were weighed and the pickers paid, the petals went next to *le tri,* or triage. Sorters sat among mountains of petals reaching over their heads, sifting through them to remove damaged petals, leaves, or any other unwanted material. The petals that made it through were

FIGURE 4. Another scene from Grasse. The figures in the background are using a press to squeeze essential oils out of plants. The one in the foreground is placing a cloth smeared with fat in a frame. He will then place flower petals on it, a process known as "enfleurage à froid."

layered upon cloth sheets that had been covered with a thin layer of solid fat. These sheets were stretched across wooden frames, or chassis, and stacked one on top of another. The process had to be done quickly, the petals deposited within a few hours of being plucked, and sometimes the whole town showed up to help. When the scent of the petals was exhausted (usually one day for jasmine, two or three days for tuberose), workers picked off the old, depleted petals and replaced them with a new layer. They repeated this process for weeks until the layer of fat was sufficiently impregnated with the flower's scent.

Working conditions were far from enlightened. Children were favored for their tiny fingers and proximity to the ground. Women earned about half as much as men for each load they brought in.[14] The amount of labor involved was stupendous. It could take an hour to pick the 4,000 tiny flowers that made up a pound of jasmine, and it ultimately took 750 pounds of flowers (or 3 million individual flowers) to produce a single pound of jasmine absolute.[15] The entire region was transformed to extract every elusive drop. Acres of vegetation were concentrated down into small vials, then sent north, into the hands of an increasingly powerful and demanding royal court.

The Perfumed Court of Versailles

The production in Grasse rose hand in hand with the power of the French crown, which conscripted perfume into its project of royal dominance. When Catherine de Medici had first arrived in France in the sixteenth century, the throne still shared power with a number of important aristocratic families. They regarded her scented Italian gloves with suspicion and spread rumors that she used perfume to cover up the smell of poison after she sent a pair to a rival, Jeanne d'Albret, who died soon after. But when Louis XIV ascended the throne a century later, he had consolidated much of the state's power in his own person. By now, scented gloves were ubiquitous and merely the tip of a much-perfumed iceberg. "Never had a man loved odors so much," wrote the duc de Saint Simon about the king. Versailles, the palace Louis XIV built as a monument to absolutism, became impregnated with them.[16]

Almost everything at Versailles was perfumed. A particularly prized item, one often exchanged as a gift between sovereigns, was a finely scented piece of fabric known as the *toilette*, from the diminutive of *toile*, the French word for cloth. It became popular to place the cloth over a table on which was arranged all the numerous products involved in the grooming routine—"la toilette," as the process itself came to be known. And nowhere did this process become more exalted and time-consuming than at Versailles.

The three central components of the toilette were pastes, powders, and pomades, all of them heavily perfumed. Pastes, or thick creams, were applied to the skin. These included *blanc,* which whitened the skin, and *rouge,* which reddened the cheeks. Powders, made from finely ground starch, absorbed the scent of fresh flower petals, with rose, musk, jonquil, and oakmoss among the popular scents. These were applied to the skin, heightening its pallor, and the hair. Rendered fats such as refined tallow or suet, thoroughly impregnated with scent, made up the pomades used to style and perfume the hair.

The regime of the toilette did not, famously, involve bathing. Water, particularly hot water, was avoided as dangerous to the health, leaving one vulnerable to disease. Yet despite his lack of baths, Louis XIV smelled good enough to earn the nickname "the sweet flowery one." He cleaned himself frequently by rubbing scented *esprit de vin* on his skin. It was also common to use vinegar, as well as soaps such as the *savonnettes de Bologne,* made with oranges, rose water, and a wide variety of other scents.[17] Specially prepared cloths, called *mouchoirs de Vénus,* made with elaborate preparations that included lemon and cloves, were rubbed on the body.

Scented linen undershirts, changed several times a day, absorbed bodily secretions. The king's laundresses steeped his in a particular preparation of *aqua angeli* that involved rose water, benzoin, jasmine, and orange-flower water. Members of the court placed sachets of flowers in their clothes and rose petals in their hair. Scented handkerchiefs and fans were de rigeur, and courtly etiquette revolved around their use. The king's physician developed a device, the *cassolette royale,* which diffused scents through the steam of boiling water. The king liked to have a different fragrance in his chambers every day until, late in life, he developed a sensitivity that made all odors unbearable except the scent of oranges from his own trees in Versailles.

The Flowery One's successor, Louis XV, only heightened emphasis on the olfactory, creating a court of extravagant redolence where even the fountains ran with perfume. Every aspect of the toilette was taken to extremes. Blanc and rouge were applied so thickly that hardly a hint of natural skin tone shone through. Large vats of pomade helped sculpt the elaborate new coiffures that piled higher and higher above the wearer's

FIGURE 5. An example of men participating in the toilette. The cloth, likely scented, is visible on the right, with pomades, brushes, and more arranged upon it.

head. Women tended to have their own hair augmented with hairpieces and powdered with tinted starch, usually gray or blue, but also possibly pink or violet. Men wore wigs, powdered as white as possible. Everything was heavily scented, and members of court paid dearly for it, hiring perfumers to develop signature scents just for them, at great cost. For Madame de Pompadour, the king's chief mistress, perfume ranked as the single greatest expense of her household.

A scented cloud enveloped those navigating the halls of Versailles, who surrounded themselves in perfume as if their lives depended on it. And, for them, it did. In an era that blamed "bad air" and unpleasant smells for the transmission of disease, Versailles became a fortress against the filth and contagion that threatened from all sides. The steam diffusers wafting scent through a room, the vinegar sprinkled about for cleanliness, the vials of aromatic water worn around the neck, the scented fans and handkerchiefs, the herbal compresses and bath soaps—far from frivolities, these were the best means available for keeping plagues and insalubrious miasmas at bay. The vast resources poured into supplying the court with perfumes can be seen as an effort to hoard the essence of life itself, to extract the vitality of Provence's hillsides and bring it north to the court.

2

The Essence of Life

Was there a way to capture the essence of life? To isolate whatever made a living thing alive, and preserve it forever in a bottle? Chemistry's first origins were found in that impulse, as those who practiced alchemy sought an elixir capable of preserving and extending life, and warding off decay and death. Although alchemy is often popularly associated with the effort to turn base metals into gold, this transformation was part of a broader project to reveal and manipulate the inner, often hidden essences of the natural world. The most coveted substance of all was the pure essence of life itself. The alchemist's task was to wrest this secret from living things, to isolate and purify the eternal and enduring from the corruptible weight of the flesh. Distillation, which separated the volatile from the inert, was a favorite approach. Plants withered and died, but certain aspects could be distilled and preserved forever. Was this a tantalizing clue to the secret of vitality and, perhaps, immortality?

The principle of distilling was simple: the volatile portion of a substance would vaporize when heated, and the vapor then traveled down a tube, where it was cooled and made to recondense, separate from the less volatile portions left behind. But in practice, it was the most recondite of arts. Each step required a delicate and individual coaxing, and inspired a tangled profusion of different designs. The first distiller on record was Maria the Jewess, a female alchemist working in Egypt around the first century CE. She was the namesake of the bain-marie (or Maria's bath), and also developed a distilling apparatus consisting of a cucurbit, or still-pot, where the substance was heated, and an ambix, or still-head, that channeled the vapor. By the tenth century, Islamic alchemists had refined the procedure, and with the addition of the Arabic article *al-*, transformed "ambix" into "alembic," now used to describe the entire apparatus. They specialized in the production of rose water, made by placing rose petals in an alembic with a small amount of water and gently heating them. Ibn Sina, known as Avicenna in the West, spoke highly of its varied medicinal uses, as well as its use in cooking.[1]

The volatile essences of fragrant plants were some of the first and most desired things to be distilled, more so than alcohol. While Islamic alchemists had noted the burning properties of distilled wine, its low-boiling vapors made it frustratingly hard to capture until the twelfth century, when several improvements centered in Italy transformed the process, such as the addition of salt and tartar to draw off more of the water.[2] Venetian glassblowers also managed to produce new, all-glass alembics that did not break upon heating. In the thirteenth century, the Florentine doctor Taddeo Alderotti had the local glassblowers make what he called a wormcooler, a long, coiling tube of blown glass. It later comes to be called a serpentine, as it often winds around the cooling trough like a coiled snake, and it proved particularly useful for the distillation of fermented liquids, like wine.

The distillate dripping from the coiled spirals of the wormcooler was a marvel. It concentrated the intoxicating properties of fermentation and stripped wine of its incidental features to isolate its essence. Clear and colorless, it looked like water, but behaved like no water ever did, burning with a blue flame usually reserved for the hottest part of fire. Because of this, Alderotti named it *aqua ardens*—burning water or fire water. He also noted its remarkable ability to preserve the essences of plants and gave recipes for concoctions of fruits, herbs, and spices distilled with it. (The word *alcohol* already existed, but it meant something else and was not used for this new substance. It was originally the Arabic transliteration of the

FIGURE 6. A sixteenth-century still, from Conrad Gessner's *The Newe Jewell of Health*. The fluid would be heated in the boiler on the right, with the volatile component traveling through the neck, passing through the cold water in the barrel, and condensing into the receiving flask on the left.

Egyptian word *kohl*, a dark gray powdered mineral used for eye makeup. By the time of the Arab alchemists it had come to mean the most fine or subtle part of something.)

It was not obvious how this remarkable substance fit within the existing natural order. The Aristotelian system, in place for 1,500 years, held that all things on Earth were made up of four elements: earth, water, air, and fire, in order of increasing subtlety and rarefaction. *Aqua ardens*, which looked like water but burned like fire, posed a paradox for the terrestrial elements. But there was another option. A fifth element, the quintessence or ether, was a substance of complete rarefaction that made up the objects in the celestial spheres. This element was not supposed to be present on Earth, but the very difficulty of isolating *aqua ardens* led many alchemists, by the fifteenth century, to identify it with the quintessence. No matter how many times they repeated the distillation, they could never completely separate out the true essence of *aqua ardens* from what they called its watery and earthy elements—a result one might expect when searching for the celestial element whose perfect subtlety made it belong to the heavens alone. The composition of the heavens offered another clue to its nature. The Sun and stars were eternal and unchanging, unlike the corruptible terrestrial realm, so full of death and decay. The quintessence, a form of matter that resisted the corruption of time, thus hinted at the secrets of eternal life.

To purify it, alchemists passed the *aqua ardens* through their alembics again and again. The goal was what they came to call *aqua vitae*—the water of life. "The name is remarkably suitable," wrote the thirteenth-century physician Arnold of Villanova, "since it is really a water of immortality. It prolongs life, clears away ill-humours, revives the heart, and maintains youth."[3] The name became widespread in common tongues: French eau-de-vie, Scandinavian aquavit, Scottish whiskey, and Slavic vodka all translate as water of life. The English *brandy* had a different source, coming from the Dutch *Brandwijn*, meaning burned wine, but the word still pointed to its origins in the fires of the alembic.

Distillation received its most thoroughgoing theoretical explanation at the hands of the Renaissance provocateur Paracelsus, who straddled the worlds of medicine, alchemy, and natural philosophy while managing to anger the traditionalists of all three. Rumored to dictate his treatises while drunk, Paracelsus was famously expansive in his theoretical approach, combining Aristotle's four elements and the *tria prima* of practicing alchemists—salt, sulfur, and mercury.[4] He then elaborated his own list of five principles that would be separated in the process of distillation: spirit, salt, oil, earth, and phlegm, which got its name from the Galenic humor containing Aristotle's "watery element." Paracelsus distinguished what he called passive principles like phlegm and earth from the active principle of spirit, which he equated with alchemical mercury as well as the Aristotelian quintessence.[5] The practical task of the distiller was to separate out all the principles, isolating the most volatile and prized of them all: spirit. It was a spiritual exercise in the most literal sense: freeing what was incorporeal and eternal from the deadened weight of its earthly body. For fragrant objects, this was the *spiritus rector*, the "guiding" or "presiding" spirit. For wine, it was the *spiritus vini*, the spirit of wine.[6] The residue left behind in the still, shed by the spirit, was known as the *caput mortuum*, or dead body.

Seeking immortality, Paracelsus died at forty-six. But his framework for distillation lived on, dominating understanding for the next 200 years. The aroma of a plant continued to be cast as a physical substance, albeit a volatile and subtle form of spirit, the "guiding spirit" as they called it, that directed matter from its brute form to the complex organization so uniquely characteristic of living things.

PERFUME AS MEDICINE

With aroma closely tied to the vitality of plants, the effort to capture and bottle it blurred the lines between perfume and medicine.[7] Reigning medical theories tightened the association by attributing most disease to "bad air" and its foul smells. In *Suspicions about Some Hidden Qualities of the Air*, the seventeenth-century chemist and natural philosopher Robert Boyle called air "a confused aggregate of effluviums" that could affect a person's health.[8] By the eighteenth century, Boyle's suspicions had been developed into a solid conviction that impure air was the root cause of disease. Living bodies, it was believed, were held together by a principle of cohesion that was continuously threatened by dissolution and decomposition, which natural philosophers described as an internal movement destroying the arrangement of the parts. These forces of putrefaction, identifiable by a change in smell, infected the air with each fetid exhalation. Breathing them in could hasten putrefaction in one's own body. Medical literature contrasted insalubrious "mephitic air," exhaled from the body, with the life-sustaining "vital air" that was inhaled. Doctors warned further of miasmas—infected air through which nearly all disease was thought to be transmitted. The only indication of these invisible infections was smell, and thus the best way to preserve health and extend life was to purge one's environment of these bad smells.

While unpleasant smells indicated the presence of the forces of decomposition and the destruction of a body's organization, the more pleasant aromas of herbs and flowers were associated with vegetative growth and the endowing of order onto an organic body. The effort to bottle these smells and sell them as curative remedies drove the trade in medicines. Distillers found that they could soak fragrant plants in wine and then distill it to produce a spirit would retain the plant's aroma. It was one of the most remarkable properties of alcohol, that it could hold on to the volatile oils that normally dissipated so quickly in the air and preserve something as ephemeral as the scent of a flower.

The reputation of spirituous medicines rose in the wake of the Black Death, although they remained rare and expensive. The practice of distilling spread slowly, with techniques and recipes carefully guarded as secrets. Monks were particularly successful at developing complex recipes that they passed down to their brothers but kept secret from outsiders. The monks of the Carthusian order used 130 different plants and flowers to produce their own "elixir of long life," Chartreuse, which one can still find in liquor stores today. The Carmelite monks at Narbonne specialized in Eau de Mélisse, which used lemon balm, lavender, and over twenty other secret ingredients, which the tightly corseted women of Versailles took to carrying in small vials to ward off faintness or the vapors.

Apothecaries, who did not share the monks' vow of poverty, began to sell medicinal extracts of herbs and bitters on a larger scale. They touted alcohol as "the mistress of all medicines" for its ability to take up plant essences.[9] By the fifteenth century, distilling guides listed recipes of elixirs, either *simplicia*, consisting of a single essence, or *composita*, consisting of several. These guides elaborated the medical benefits, which covered nearly every ailment known to man, from baldness to dropsy, from the bite of a mad dog to excessive farting.

A favorite recipe was Queen of Hungary water, made from spirits distilled with rosemary. Prescribed by physicians, it could be smelled, drunk, or rubbed on the body to treat a range of ailments from headaches to colics, or simply to promote general vitality. Later recipes often called for the addition of lavender, bergamot, jasmine, and other sweet-smelling

FIGURE 7. The back room of an eighteenth-century apothecary, containing a number of alembics and distilling flasks. The front of the shop is visible through the doorway.

florals that left it as much prized for its scent as for its therapeutic value.[10] Even more widespread, if somewhat more pungent, was vulnerary water, sometimes known as Eau d'Arquebusade from the fifteenth-century musket whose wounds it was intended to heal. But while the monks' original intent was to reduce scarring, infection, and gangrene in violent wounds, the water took on a variety of uses and was often gargled to freshen the breath. Recipes varied and could have upwards of seventy-five different botanicals, but most tended to include sage, angelica, absinthe, and hyssop.

Apothecaries had another prized substance, vinegar, which shared with fats and alcohol the remarkable ability to draw out the fragrant essences of plants. Getting its name from the term *vin aigre,* or sour wine, the substance had been known since antiquity as the unfortunate consequence of leaving wine too long exposed to air. But by the fourteenth century, physicians and alchemists had incorporated it in their *materia medica*, concocting elaborate recipes infusing it with botanical essences. A favorite was the vinegar of the four thieves, which took its name from the story of a group of bandits apprehended looting the houses of plague victims. The first question the magistrates had for them was how they kept from contracting the plague themselves. They escaped execution only after revealing their secret recipe, a complicated affair of some dozen ingredients, including cloves, wormwood, juniper, and camphor.

France started regulating distillers in 1624, when they were first grouped together with the apothecaries and spicers. By 1639, they formed their own guild, with their masters earning the title "distillers of eau de vie, strong waters, oils, essences, and spirits."[11] The making of vinegar was originally under control of the apothecaries, then distillers. Vinegar makers eventually formed their own corporation. Antoine Maille, the official supplier to the court of Louis XV, was listed as a "distiller vinegar-maker." The shop he opened on the rue Saint-André-des-Arts in 1747 sold 180 different types of aromatic vinegars, which could be smelled, drunk, applied to the skin, or used to preserve foods.[12] It was one of the chief techniques for purifying air, whether sprinkled around a room to combat odors or held in a cloth up to the face.

By the time Laugier moved to Paris in the 1770s, the city's thriving apothecaries and distillers had come to rival Grasse in the market for perfume. In Grasse, the glovers had firm control over the field of perfumery, but in Paris the field was much more porous, where no single corporation retained control of the perfume trade. Glovers vied for control with apothecaries, spicers, and haberdashers, and not always successfully. Indeed, in the sixteenth century glovers in Paris were forbidden to sell any perfumes they had not made themselves, as a way to keep them out of the apothecaries' market. The 1656 statutes loosened this requirement, but apothecaries retained an upper hand in the city, which by 1725 had become one of the important perfume centers of the world, hosting more than four times as many perfume sellers as Grasse.

Newly arrived in Paris, Blaise Laugier had thrown his lot in with the apothecary tradition, showing up first on the list of the city's apothecaries and later marking his profession as "perfumer-distiller."[13] He had an alembic in the back and carried all of the standard cures: vulnerary water, Eau de Mélisse, Queen of Hungary water, and a wide number of elixirs, spiritous *eaux*, and quintessences of various botanicals. He sold Four Thieves vinegar as well as an "anti-pestilential" of his own design which promised to "chase away the bad air."[14] Indeed, he sold up to seventy-nine different kinds of vinegars, which he divided into different categories: There were those for "the bath," made with lavender, thyme, laurel, and other herbs, ones for "la toilette," made with rose, jasmine, and other florals, sold in smaller vials. There were those "for the table," meant to be consumed, that were infused with everything from berries to truffles to anchovies. (Of these consumables, cloves and cinnamon were the most expensive.) Then there were those "for cleanliness" that, when applied to the skin, promised to cure wrinkles, alleviate freckles, soothe razor burn, whiten skin, and remove blemishes, corns, and pimples.

And where did Blaise learn these techniques, so different from the straightforward steam distillation common in Grasse? He followed, he said, the instructions of the apothecary Antoine Baumé, who was making public the long-held secrets of his craft.

THE PHILOSOPHICAL SPIRIT

PARIS, 1760S

The arts of perfumery and distilling were both born from alchemy, a practice so secretive its followers communicated in a code impenetrable to the uninitiated. But Blaise Laugier had arrived in Paris at a moment when everything was changing. The spirit of the Enlightenment sought to shine light in the dark corners of human knowledge and make public every secret hidden away. Diderot had just begun to publish the first volumes of the *Encyclopédie* and, in common cause, two fellow Parisians, Antoine Baumé and Pierre-Joseph Macquer, had teamed up to teach a public course on chemistry that laid bare the mysteries of distillation.[15] Genuine knowledge, they claimed, had to be both theoretical and practical. So while Macquer, whose strong features and cleft chin brought to mind Rousseau, lectured on the theory, Baumé, whose impish air and pointed nose gave him an uncanny resemblance to Voltaire, ran the demonstrations, performing over 2,000 experiments in front of an audience drawn from across the whole of Paris.[16]

Macquer, the theoretical one, sought to update Paracelsus's account. Fashioning himself the Euclid of chemistry, he built a philosophical system that went from the simple axioms of Aristotle's four elements to the complex products of spirit and oil. He defined spirit, or *esprit,* as all the liqueurs removed from different substances by distillation. There were three kinds: flammable spirits, acidic spirits, and alkaline spirits. The flammable spirits were the most interesting and most volatile. They could be further broken down into two categories: the *esprits ardent,* drawn from wine, beer, and other fermented liqueurs, and the *esprit recteur,* drawn from the fragrant essential oils of plants.

For Macquer, the core battle of distillation was the one between spirit (the volatile, life-sustaining part) and phlegm (the corruptible element that befouled it). Liberating the spirit from the phlegm was not easy. The difference in volatility was often subtle, and a large amount of what the distillers did not want often accompanied the desired spirit through the alembic.

As both essential oils and alcohol were flammable, it was clear that they also contained phlogiston, an eighteenth-century addition to the list of chemical elements. Macquer, following the German chemist Georg Stahl, defined phlogiston as the flammable principle, present in any substance that could catch fire and absent in any that could not. It was not, he admitted, well understood, but by the 1760s, he was leaning toward the position that the spirit of wine was phlogiston itself, combined with water. For the chemists, phlogiston displaced the alchemical philosopher's stone and played a central role in the processes of life and decay, a by-product of both exhalation and putrefaction. Macquer emphasized its role in the production of mephitic gas, a putrid air emitting from decay that was unable to support life or combustion. It was on the basis of this work that the city of Paris decided to close and exhume the Holy Innocents Cemetery after reports that the neighboring inhabitants found their candles extinguished when they entered their cellars, which were heavy with the mephitic air.

This was the theory. In practice, no one had ever isolated any these substances: phlogiston, *esprit recteur,* or *esprit ardent.* Their extreme volatility meant they immediately dissipated into the air. And the distillation techniques used to produce them were still far from perfect. Baumé, the practical one, took up the challenge. "Useless and awkward," was his verdict on the existing equipment.[17]

Baumé admitted that *esprit recteur,* responsible for aroma, was impossible to isolate, but he gave several practical tips on how to know its properties. There was a particular flower, the fraxinella, which impregnated the air with so heavy a scent that it could catch fire—a brief, volatile flash that left the plant unscathed but odorless. This was proof, for Baumé, that these fragrant exhalations consisted of a flammable vapor that Baumé called "the ethereal liqueur of vegetation." The invisible vapor could also be detected in steam distillation of lavender or thyme, as some perfumers learned the hard way: the expansive vapor was the first material to pass through the still and could explode the condenser if it was sealed too tightly.

But most of Baumé's attention was directed toward the *esprit ardent* and, in particular, the making of brandy, or eau-de-vie. The drink had begun to gain a foothold in Paris by the eighteenth century. Brandy sellers now roamed the streets of Paris, selling drinks from a wicker basket full of bottles and glasses slung from their neck. But it was a poor product they sold, generally made from wine "that one cannot sell because of the bad quality" and drunk only by "soldiers and the common folk."[18] As Mercier put it in his description of working-class Paris, "porters and peasants toss down this liquor, the soberer of them drink wine."[19]

Baumé blamed outdated distillation techniques for this "bad eau de vie" with its "disagreeable odor." Too many distillers continued to use the same alembics passed down to them "from time immemorial" without

FIGURE 8. A woman selling brandy on the streets of Paris, 1737. The caption has her calling out "La vie! La vie!," or "Life! Life!"

any thought to the theory governing the process. He listed a number of improvements to ensure their product did not "contract an odor," emphasizing that one must never boil the wine over an open flame.[20] He took care to distinguish between the eau-de-vie sold on the streets and the purer *esprit de vin* that had been "rectified" or passed through an alembic multiple times. In his 1762 textbook, *Theoretical and Practical Elements of Pharmacy*, he stressed that pharmacists should only use rectified *esprit de vin* in their recipes, providing step-by-step instructions in his chapter titled "Tinctures, Elixirs, Quintessences, and Spirituous Balms."[21]

Baumé's most explicit effort to publicize the secrets of distillation came in 1777, when he won a prize competition held by the Society for the Encouragement of Arts, Crafts, and Useful Inventions. They had offered 1,200 livres for the best essay on the question, "What are the most advantageous forms of the stills, furnaces, and all the instruments used in the work of the large distilleries?" Baumé responded with detailed instructions on six new alembic designs, which the Society published and distributed widely. In his response Baumé pointed out that Paris was poised to become the center of high-quality distilling, as it was only "in the big cities where the savants and the artisans are united, that one can hope to perfect an art so useful to commerce."[22]

The purer spirits of these improved stills made excellent carriers for perfumes. By the time Marie Antoinette became queen of France in 1774, tastes in perfume were shifting from scented leathers and fat-based pomades to spirituous *eaux de toilette* and vinaigrettes. Versailles remained as extravagantly scented as ever. "What a debauchery of jewelry and of perfume," proclaimed the Swede Axel von Fersen when he first arrived at the court of Louis XVI and Marie Antoinette.[23] The queen went through eighteen pairs of perfumed gloves a week, and commanded more powder and pomade than ever for her signature pouf that rose several feet high. But the heavy "animalic" scents popular in the seventeenth century, such as musk, civet, and ambergris, had lost ground to lighter "vegetable" scents of florals, citrus, and woods. The perfumer Fargeon catered to the queen's demand for "naturalness" by creating a new perfume exclusively

FIGURE 9. Antoine Baumé's still from his textbook, *Elémens de pharmacie théorique et pratique.*

of the delicate, floral notes of orange blossom, bergamot, lavender, galbanum, iris, violet, jasmine, jonquil, and tuberose. He named it Parfum du Trianon, after the gardens where the queen would escape to playact living the country life. Her return to nature, however, had its limit, as the sheep, goats, and cows she tended for amusement were all doused in perfume to hide their animal scent.

PARIS, 1775

Even the Paris Academy of Sciences was involved in ensuring that French cosmetics were healthy and natural. In 1775, they commissioned a study of the rouge used to color women's cheeks. This staple of the courtly toilette had traditionally been prepared with a combination of cinnabar (a toxic form of mercury) and lead (also toxic), and complaints that it blackened the skin and sickened its users mounted. The Academy sent one of its youngest members, Antoine Lavoisier, to visit some dozen perfumers of Paris and collect samples of the rouge they sold. He found that while the mineral version of the cosmetic had toxic effects, some sellers provided a version made from plants, such as saffron or safflower, that was vastly preferable.[24] Laugier was among those selling the healthier version, and a subsequent article in *Le Mercure* singled out "his beautiful vegetable rouge" as part of "the most complete success" of Laugier's business.[25]

This was far from the last commission of Lavoisier, an exceptionally ambitious and serious young man from a well-connected family. His father was a prominent lawyer who wanted his son to follow in his footsteps. Lavoisier had obligingly gone to Paris to study law, but while there he developed a furtive passion for chemistry, attending chemistry lectures when he could and reading Macquer's *Dictionary of Chemistry* in his spare time. Although he passed the bar, he decided not to practice law, and instead became a member of the Academy of Sciences that same year.

To support himself while doing chemistry, he took a gamble on a uniquely French institution known as the General Farm. In France, the government did not collect taxes itself, at least not the "indirect taxes" on products such as salt, tobacco, and alcohol that made up nearly half of the state revenue. Instead, the job fell to a private organization of sixty men, known as the General Farm. They paid the state a sum of money in advance, and then kept as profit the surplus left over after they had finished collecting, an amount that was hard to know in advance but was always large. Lavoisier had bought a position on the Farm at the age of twenty-five. The post did require some work, primarily as an inspector charged with detecting fraud and the overseer in Paris for the taxes known as *aides* on tobacco, alcohol, playing cards, meats, oils, and soaps. But, as he had hoped, it left plenty of time for him to pursue his own interests.

He had picked up another job as well, as director of the Royal Gunpowder and Saltpeter Administration. After watching France lose the Seven Years War by running out of gunpowder from its private enterprise supplier, he had convinced the king to take control of the country's gunpowder supply and soon found himself running it.[26] He moved onto the grounds of the Petit Arsenal with his new wife, Marie-Anne Paulze, in 1776. He had somehow managed to find a family even wealthier than his, marrying the seventeen-year-old daughter of a fellow farmer general. In the attic of the building, he built the finest chemical laboratory in Paris. He rose at 5 a.m. each morning, mounted the steps to the laboratory, and proceeded to divide the day methodically between his own research, his duties to the tax farm, and his work on gunpowder. By working until 10 at night, he carved out enough time to spend at least six hours a day on chemistry.[27]

Lavoisier soon became ridiculously, fabulously wealthy, bringing in $48 million in tax revenue between 1768 and 1786. "M. Lavoisier," a journalist wrote, "unlike his fellow chemists, did not have to seek the philosopher's stone, he simply found it in his office."[28] He applied his chemical knowledge to tax collecting. France had, from 1687, designated different categories for the selling of brandy, depending on the percentage of spirit it contained. Sellers had to specify whether they were selling eau-de-vie *simple* (with the least amount of spirit), eau-de-vie *double,* or *esprit de vin* (which had the most). Each of these categories was, in theory, taxed at a different rate.

In practice, it was hard to distinguish between them. Distillers generally relied on their own evaluation of a product's taste and odor to determine its strength, although there were a few additional tests. The most important of these was to shake the brandy in a special bottle called an *épreuve* (or proof) and examine the resulting bubbles. If they disappeared quickly, it was spirit strength. If a there was a diffuse foam, there was too much phlegm. The sweet spot for eau-de-vie was when there were exactly three bubbles left, known as the *trois perles*.[29] For the stronger *esprit de vin*, distillers checked that there was no residue after it had been set on fire, or, more dramatically, they mixed it with gunpowder—if the gunpowder ignited, the *esprit* was strong enough. But these tests were more art than science. Without a reliable way to distinguish between them, all brandies were taxed at the same, lowest rate.

Here is where chemistry became useful. In addition to improving distillation, Baumé had invented a device, the hydrometer, that could measure the percentage of alcohol in a solution of alcohol and water based on the differences in their specific gravities. Using his scale, manufacturers could assign a number to mark the "proof" of their product and distinguish between the various grades. (*Esprit de vin*, for example, was defined as being above 80 percent alcohol, or 160 proof.) Lavoisier quickly mastered the use of a hydrometer. In 1770, the taxation at different rates kicked in, in Paris only, making highly distilled spirits more expensive there.

Blaise Laugier, who traded in the highest proofs possible, felt the pain. He tried to get around the tax laws in 1779 by buying property in Chantilly, just outside Paris. The town was under the jurisdiction of the Prince of Condé, who had worked out an arrangement to avoid paying taxes entirely on eau-de-vie, as long as it was locally consumed. Laugier set up operations as a starcher, making fine powders for the hair and body, although the town records indicate that he "extends his commerce to various other objects, such as scents and soaps, etc." He could then sell his distilled products tax-free. But by 1780, finance minister Jacques Necker was getting suspicious. Officials had counted 3,325 veltes (roughly 6,000 gallons) of eau-de-vie coming into the small town of 1,500 inhabitants in a single four-month period and concluded it could not all be for personal consumption.[30] The tax haven closed, and Laugier sold his starch factory in 1785 and returned to Paris full time.

The smuggling of distilled spirits, far from stopping, was fast becoming a distinctly Parisian art form. Underground pipes, secret doors, and elaborate ruses all circumvented the customs officers stationed at the gates of the city. Lavoisier, crunching the numbers, determined that smugglers were getting one-fifth of their merchandise through, tax-free. To stop them, he proposed building a vast wall around the city of Paris, fifteen feet high, whose sole entrances would consist of elaborate tollbooths where everyone would be subject to a thorough inspection. Construction began in 1787, outraging the city's inhabitants, who complained about being imprisoned in their city, and about the "unhealthy air" resulting from the enclosure.[31] Worse yet, the wall cost millions of pounds, and the money collected went straight into Lavoisier's pockets. The joke around town was that, as a chemist, "he wanted to put Paris in a cucurbit whose receptacle would be the Farmer's cashbox."[32]

But Lavoisier did not shy away from controversy. That same year, he published a work overturning the accepted principles of chemistry. By this point, the Aristotelian system, with its four elements, had been in place for 1,500 years. Even Macquer and Baumé had retained the general framework, casting the element phlogiston as a substance of extreme subtlety, responsible for fire. They were, moreover, both committed vitalists, convinced that a particular life force animated organic beings, and that phlogiston played a role in the process. Lavoisier had grown suspicious of these claims early on, pointing out that if phlogiston was a substance, then it should have a weight. But he found that when mercury underwent calcination, in which it supposedly lost its phlogiston, it weighed *more* than it had before.

Aristotle's so-called elements had little place in Lavoisier's new system. Air, for Lavoisier, was not a single element but was composed of several different components, including a "vital air" that was defined by its "oxigene principle," soon to be shortened to oxygen. Water, he showed, was not an element either, but a combination of oxygen and an "inflammable aqueous principle" (soon identified as hydrogen). He took on fire as well, showing that combustion was the reaction of oxygen with a metal or organic substance, and not any kind of element at all.

A revolution in chemistry had begun. In 1787, Lavoisier, with several colleagues, published a reformed nomenclature. Among those colleagues was a young chemist named Antoine-François Fourcroy, who he had taken on as a protégé. Lavoisier complained that the alchemists used an "enigmatic language" deliberately intended to obscure meaning, where words meant one thing to the adepts and something else entirely to the uninitiated.[33] The oil, mercury, and water of the alchemists had nothing to do with oil, mercury, or water as they were commonly understood. The language of distillation was particularly bad. Lavoisier cringed at the term *pelican*—a vessel used for recirculating condensed vapors that got its name from the legend of the pelican plucking its own breast to feed blood to its young, and at the use of the skull for the *caput mortuum*, the residue left behind after distillation.

Old staples of the alchemists were transformed. *Aqua regia*, or royal water, became nitro-muriatic acid. *Aqua fortis*, or strong water, became nitric acid. Oil of vitriol became sulfuric acid. The spirit of Venus became acetic acid. Lavoisier got rid of the "butters" that were not butter (they became sublimated muriates) and the "flowers" that were not flowers (they became sublimated oxides).

Some of the key concepts of Macquer and Baumé's system did not escape the new nomenclature's knife. Phlogiston was gone, referred to only as "the hypothetical principle of Stahl." As for *esprit recteur*, the authors lamented that "we did not think we could let it survive," and substituted the word *aroma* instead.[34] *Esprit ardent*, or *aqua ardens*, became "alcohol," as Lavoisier repurposed the old Arabic word. (Both Macquer and Baumé used the word *alcohol* in its original sense, that is, to mean any essence extracted by the processes of distillation or sublimation, including dry powders.)

Alcohol was of particular interest to Lavoisier. It had replaced *spirit* in the nomenclature, and Lavoisier was intent on giving this age-old category a rigorous explanation equal to those he had given air, water, and fire. He turned his attention to the process of fermentation, which he called the "most striking and extraordinary operations of all those that chemistry presents us."[35] Macquer, Baumé, and a long line of alchemists would have agreed, seeing in the process the guidance of a living force. But Lavoisier hoped to explain it as a straightforward reaction of sugar, just as he had successfully explained combustion as a reaction of oxygen, and leave behind the vital principles of the alchemists just as he had left behind the flammable principle of phlogiston.

But banishing spirits proved frustratingly difficult. After years of work and countless repeated experiments, he admitted that he could not get fermentation to occur as a pure chemical reaction with sugar alone. The process only happened when he added a ferment—some kind of living substance, such as beer yeast. He published his results in February 1789, acknowledging his failure (a rare thing in a career in which successes had accumulated so readily). Yet, not one to give up easily, he ordered a new fermentation apparatus, hoping that greater precision would solve the matter.

The new apparatus arrived in June 1789, but he never got around to using it.[36] The atmosphere of Paris had grown too tense to ignore. France stood locked in a political crisis. The Estates General had been meeting in Versailles since May, and the deputies of the Third Estate had just broken off to declare themselves a National Assembly, vowing not to disband until France had a constitution. The king had refused, eventually calling in the army for support. As June stretched into July, he stationed seventeen regiments of Swiss and German mercenaries around Paris. They looked poised to invade, and it was clear that the people would not greet them peacefully.

Many of Lavoisier's scientific colleagues had left the city, sensing trouble. But Lavoisier's position as head of the Gunpowder Administration placed him in a difficult situation, sitting literally on top of the city's main supply of gunpowder, which was housed at the Arsenal. The Arsenal itself had few defenses, but it sat adjacent to an old fortress, the Bastille,

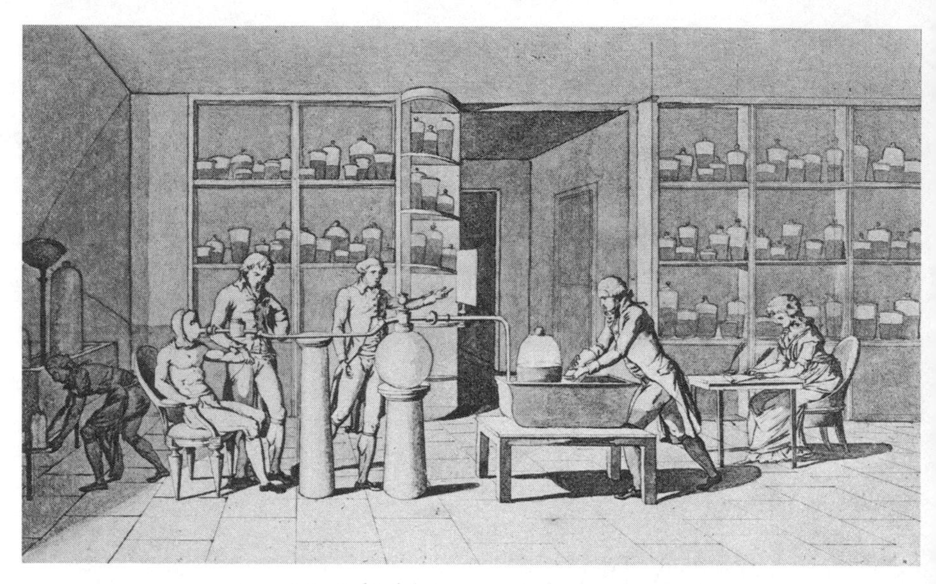

FIGURE 10. Lavoisier in his laboratory conducting an experiment on the respiration of an assistant, Armand Séguin. The illustration is by his wife, Marie-Anne Lavoisier, depicted on the right.

which was designed to be impregnable. It had been serving lately as a prison, but its commander, the Marquis de Launay, had been emptying it of prisoners and filling it with the king's troops in anticipation of making it a holdout of royal power in any unrest. As tensions in Paris increased, de Launay ordered Lavoisier to move the Arsenal's gunpowder stores to the Bastille. The operation took place just before midnight on July 12, under the cover of a waning moon. Lavoisier oversaw the hurried transfer of 250 barrels of gunpowder—21,000 pounds—into the Bastille's kitchens.

Violence broke out the next day. The first object of the angry mob's wrath? Lavoisier's controversial tax-collection wall. Angry crowds of men and women hacked at the hated wall and torched the recently completed toll booths, creating a ring of fire around the city that burned throughout the night. Before dawn broke the next morning, de Launay asked for Lavoisier's help moving the gunpowder from the Bastille's kitchen to a

FIGURE 11. The burning of the Barrière de la Conference on 12 July 1789. Parisians burned down several of Lavoisier's new toll booths, one of the first acts of violence of the French Revolution.

more secure spot in the cellars. Lavoisier sent an assistant, Clouet, who returned around 10 a.m. reporting that there was a growing crowd around the fortress. They were armed with rifles, pikes, and swords, and demanding its surrender.

From his home in the Arsenal, Lavoisier would have been able to hear the sounds of the crowd, grown to nearly 1,000 people. The gunshots began around 1:30 p.m., as scattered fighting broke out between the revolutionaries and royal forces. By 3:30 p.m., the deep boom of cannons announced the crowd's efforts to batter the fortress gates. And, finally, cheers and celebration erupted as the drawbridge dropped and the crowds stormed in. They dragged de Launay outside, severed his head from his body, and mounted it on a pike as a message to the enemies of the Revolution. Could Lavoisier see it from his window? And did he fear the crowd would turn on him? Gossip had circulated among them about his role in moving the gunpowder, and questions had arisen about his loyalties.[37] But he had just led a revolution of his own, toppling a theory of matter that had stood longer than the French throne, and he stepped up to embrace this one, as well.

3

Revolution

PARIS, TUILERIES PALACE, JUNE 1791

By the spring of 1791, the royal family had been wrenched from Versailles, dragged from their bedrooms by a mob of angry women. They moved to the Tuileries Palace, where the new revolutionary government could keep an eye on them. The king had pretended to go along with the new demands for a constitution, but in secret he was formulating a plan to escape from France and return at the head of Austria's army to reclaim his old position. Everything proceeded smoothly at first, in the strictest secrecy and without suspicion. But there was one complication: the queen refused to leave without her stores of perfume.

Marie Antoinette insisted on bringing a large trunk, known as her *necessaire,* which contained all the supposedly indispensable objects for her toilette. There were crystal bottles of perfume, labelled tinctures, oils, scented waters, liqueurs, and one marked *gouttes anodines de Sydenham,* which referred to an opium-based pain medicine and raises the question of whether jokes about Marie Antoinette's addiction to perfume had a kernel of truth to them. There were silver jars of cosmetics, a small handbell for ringing servants, and a "spirit burner" that burned alcohol to heat coffee or chocolate. She had it packed and ready to go from the time of the first insurrection in 1789. As the plan for escape took shape, she wanted to send it on ahead, across the border. Hoping to avoid suspicion, she concocted a plan to fool anyone watching. She would claim that her sister, Christina, had asked her for a vanity case of her own, and made a public show of ordering a duplicate one to be made that she could send on. But as the day of escape approached, the duplicate was still not ready, and the manufacturer informed her he needed two months more. The queen, refusing to abandon her case, had the original packed and sent, visiting perfumers on rue Saint-Honoré to fill bottles with her favorite scents. She hoped that the story about her sister was sufficient to throw everyone off her trail. It was not. The servant who helped with the packing went immediately to the mayor of Paris, raising the alarm.

The royal family was captured in Varennes, thirty miles from the border, and accompanied back to Paris by the National Guard. They were reinstalled in the Tuileries, but everything was different. The king and queen had shown their hands as enemies of the state, setting in motion the events that would lead to their trial for treason and eventual execution. The courtiers that had surrounded them fled if they could, leaving France with their own hastily packed bags.

The political power of France now resided in a repurposed horse arena where the Royal Equestrian Academy had practiced in calmer times. The National Assembly met there now, writing a new constitution for France in the only room large enough to fit its roughly 1,200 members. It was a far cry from the grandeur of Versailles ("meager and uncomfortable" was Maximilien Robespierre's assessment).[1] But the Assembly spared no expense in their efforts to purify the air and "to banish from the hall the bad odors and putrid exhalations."[2] They burned aromatics in the great hall, evaporated fragrant vinegars several times a day, and used *esprit de vin* to clean the carpets. They even had someone disperse scents in the air during sessions. "Messieurs, it is a question of health," the deputies stated, "of the life itself of the representatives of the nation, and of all the citizens whose patriotism leads them to our sessions."

Neither did they intend to restrict these procedures to themselves alone. Rather, they sought to bring the products of health and vitality, previously in the hands of so few, to the entire population. They blamed the existing inequality on the twin evils of hoarding and monopolies, which perverted the market and created artificial scarcity. To combat it, one of their first orders of business was to abolish the guild system, which they equated with monopolies, existing only to restrict the number of people who could practice a trade. The National Assembly included the medical schools in their ban, condemning them as the most tightly controlled professional monopoly of all.[3] The Faculties of Medicine had required years of study to memorize classical texts in Latin and debate how to best balance the Galenic humors. The number of doctors produced was so small that most people in France had no hope of ever consulting one, leaving as their only option the apothecaries, grocers, and perfumers who prepared their compounds. To make things worse, the physicians actively prosecuted anyone else practicing medicine, in the name of combating charlatanism. It was a system designed to allow the nobility and royal court to hoard access to doctors the way they hoarded everything else. To dismantle it, the Assembly shut down the medical schools and outlawed the title of "doctor." They replaced it with the more inclusive *"officier de santé,"* which dropped the traditional distinctions between physicians, surgeons, and apothecaries.[4]

What would health care look like if available to all citizens? The National Assembly appointed Joseph Ignace Guillotin to work it out. (Guillotin had, by the way, already conceived of the design for his namesake decapitation device but had not yet ordered its construction.) He headed a new Comité de Salubrité, or Health Committee, which shifted the focus of medicine from the physician's old system of Galenic humors to the effort to maintain clean and hygienic environments.[5] It sought to bring health care into the universe of basic human rights, and emphasized that this brought with it a corresponding set of reciprocal obligations. It became the citizens' responsibility to keep themselves clean and in good health and to maintain a hygienic environment. The Revolution made a civic virtue out of hygiene and a moral duty out of smelling good.

French citizens embraced the new hygiene regime, cleaning themselves as never before. Gone were the days when people feared letting water touch their skin. Although there was still no running water in houses, it became common to wash one's hands and face out of a basin. Public baths popped up on the banks of the Seine, pumping river water into tubs in private cabins. These bathhouses were unheard of when Blaise Laugier first arrived in the 1770s, but there were at least 150 by 1790.[6] Advertisements touted their therapeutic effects against "rheumatic pains, paralysis, wrenches, sprains, swellings, back pains, gout, sciatica."[7] Bathers were joined on the banks of the Seine by teams of laundresses, hired to collect the dirty laundry from households and scrub them in the river. All classes of people sent their clothes to be cleaned, at least the bottom layer, known as "linens," which medical opinion recommended changing every two to three days (and perhaps more often in the summer).[8]

PARIS, PLUVIÔSE, YEAR I (FEBRUARY 1793)

But citizens' efforts to stay clean soon ran into trouble. By February 1793, or Year I of the new revolutionary calendar, soap had all but vanished from the shelves. Many staples, like bread and wine, were in short supply, but none had disappeared more dramatically than soap. It was usually women who bought it for their households, and they scoured the city for it with little luck. Sometimes rumors spread that it could be found at a certain establishment, and everyone would flock there at once, only to find that the price had been ratcheted up beyond reach, leaving them congregating empty-handed outside, grumbling at the injustice of price gouging.[9] Washing women were particularly hard hit, and on one Saturday, 23 February 1793, a large crowd of them forced their way into the riding hall to complain to the National Convention (which had replaced the National Assembly when France declared itself a republic). The Convention's members were debating some detail of the constitution, but the women shouted over them to be heard, condensing their grievances into a single cry: "bread and soap."[10] Once they had the floor, they praised the Revolution and the efforts of the National Convention, but they complained that hoarding and speculation had driven up the price of bread and soap, and soap was impossible to procure. Their demand was simple: institute the death penalty for anyone caught hoarding soap. The next week they took matters into their own hands. Several thousand women took to the streets, more than had participated in the famous women's march on Versailles. They swarmed shops and barges, taking the soap and redistributing it at what they deemed fair prices.[11]

The situation only grew worse in the summer, as heat ripened every odor. A barge laden with soap had arrived in Paris on the Seine, but rumors began to swirl that it intended to leave for Rouen without unloading. On 25 June, a "large group of women" descended upon it, commandeered its contents, and distributed the soap at twenty sous a pound.[12] The next day, even more women showed up, overwhelming barges up and down the Seine and claiming the shipments for themselves. At the Port du Louvre, a landing on the Seine near the Tuileries Palace, the mayor donned his official sash and went down to the docks, even wading through the water to get to the scene. He spent two hours pleading with the women to stop, but they continued to unload cases of soap. The National Guard showed little interest in helping, and in the end the only group capable of restoring calm were the Revolutionary Republican Women.[13] This rather fearsome lot, known to dress as Amazons, or in soldier's helmets or police attire, was one of the most relentless critics of price gouging. On another occasion, the group marched into the Jacobin Club to complain that "the speculators, the hoarders, and the egotistical merchants" were killing people with their high prices. "Exterminate all these scoundrels," was the solution they proposed.[14]

The National Convention was well aware of the problem. They had been fielding complaints from all over France about the "extreme famine" of soap, which could be found nowhere and led to "the alteration of their health and deterioration of their clothing."[15] But it was part of a larger set of issues they were dealing with. Within days of the king's execution on 21 January, nearly all of Europe had joined in a coalition against France, leaving it surrounded by enemies with its borders shut down. The National Convention had gambled that France's strong economy could stand on its own, but soap proved to have a fatal break in its supply chain. The French had relied on Spain as the sole source of barilla, a seaside plant that when burned made an alkaline soda ash that produced the hardest, best-quality soaps. For a long time, France had imported several million pounds of soda ash every year, never able to establish its own supply because Spain had been so protective of this trade, even instituting a law that made smuggling barilla seeds out of the country a crime punishable by death.[16] In a pinch one could make soap from another alkali, potash, made from the ashes of hardwood trees. But these were also in short supply in France, which had been getting its potash from the United States. The French people had hoped that their "sister Republic" would continue trading with them, given that they were not at war with each other. But pressure from Britain cut off the US supply line as well.

And there was another issue. Alkali was necessary to make not only soap but gunpowder, and that was a bad thing to run out of in the middle of a war against the entirety of Europe. It was a stark decision straight out of a morality play. In this moment of crisis, the National Convention urged its population to stay healthy and to fight for their country. But it scarcely had enough ashes to support one, let alone both tasks. It turned to the chemists for help. It had been a chemist, Lavoisier, who figured out that treating saltpeter with alkali produced not only more gunpowder, but stronger stuff that increased the range of their cannons by over 50 percent. Lavoisier had bragged that his gunpowder had won the American Revolution, and the National Convention hoped that a similar application of chemical knowledge could get them out of their bind.[17]

Lavoisier, however, was out. He had continued to struggle with the trust of the sans-culottes. In August 1789, a crowd surrounded a barge docked on the Seine, convinced that it was filled with arms intended to suppress the Revolution. Lavoisier had approved the export of large amounts of an inferior grade of powder known as *poudre de traite*, or slave-trader's powder, to make room for the arrival of new shipments. But the semiliterate crowd, misreading the words as *poudre de traitre*, or traitor's powder, assumed Lavoisier had conveniently labeled the barrels to match his nefarious intentions. They dragged him through the streets, many shouting for his immediate hanging at the Place de Grève. They ended instead at the Hotel de Ville, where he persuaded them of his innocence in an impromptu public trial. Lavoisier seemed more exasperated than afraid. "It's about time," he wrote after yet another incident where the town of Étampes blocked a shipment, "that somebody convinced the town councils that gunpowder is not in the hands of the nation's enemies, but that, on the contrary, those who are in charge of producing and distributing it are second to none in their patriotism."[18]

As the Revolution had grown more violent, he had grown less supportive. In September 1792, when armed mobs began pulling aristocrats from prisons and massacring them, he fled from his home in the Arsenal in the middle of the night, resigning his position at the Gunpowder Administration. The National Convention replaced him with Antoine-François Fourcroy, his friend and closest collaborator, who had coauthored the *Nomenclature* in 1789 and helped invent the new system of chemistry.

Fourcroy had worked his way to scientific prominence from humble beginnings, living for years in a small attic apartment with a ceiling so low that he could only stand up if he kept his head tilted at an angle. His father had worked as an apothecary before he was born, but he was forced out of the job when the guild tightened its requirements, and he struggled to feed his family. Fourcroy himself had tried to go into medicine but was rejected by the physician's guild for political reasons. Instead, he started working in Lavoisier's laboratory at the Arsenal, doing various odd jobs. He became one of the first and most loyal converts to Lavoisier's revolutionary ideas.

Fourcroy was considerably more personable than Lavoisier, with more patience for explaining his ideas to the uninitiated, and he soon became something of the public face of the movement. He had a commanding, almost mesmerizing presence. Speaking in front of an audience, he said, felt like going into a trance. He was much in demand to give lectures around town, and wherever he went he was followed by a group of loyal devotees. Over a hundred of his fans wrote to the Jardin des Plantes when Macquer died in 1784, demanding that Fourcroy replace him. He did, and the crowds were so thick that the school had to double the size of its amphitheater, and then, a few years later, tear it down to enlarge it again, as 1,200 seats were still not enough to satisfy the crowds clamoring to hear him reveal the new chemistry. He was fast becoming the quickest way to learn about Lavoisier's revolutionary ideas. Nicknamed "the apostle of the new chemistry" as a jab, he took it as a compliment.[19]

He was at the peak of his popularity when the Revolution broke out and hoped to avoid getting too involved in politics. He refused the position at the Gunpowder Administration at least twice, submitting instead the name of his assistant, Nicolas-Louis Vauquelin.[20] The executive council would have none of it, however, and appointed him anyway, saying, "with the fatherland in peril, all citizens belong to it."[21] He was soon roped in even further. In the elections for a new National Convention, he found his name on the ballot, against his own wishes, to be a deputy from his district in Paris.[22] Also on the ballot, running against him, was a man who had been a rival for many years: Jean-Paul Marat.

Marat and Fourcroy shared similar origins. They both came from modest backgrounds, wanted to go into medicine, and were rudely rebuffed by the physician's guild. Their paths, however, had diverged. While Fourcroy went to work with Lavoisier, Marat had left for England, returning to France after several years claiming to have degrees from universities just vague enough to be unverifiable. He set up a medical practice in Paris in 1776 and soon gained a reputation as "the doctor of incurables," promoting his own idiosyncratic cures, including an *eau anti-pulmonique* that he claimed cured consumption, as well as various other medicinal waters, unguents, and chemicals.[23] He criticized traditional physicians as out-of-touch elites, and as Lavoisier gained in popularity, he attacked him, too, as another member of the elite. He wrote a book, *Les Charlatans modernes,* in 1791 that blended an attack on Lavoisier's discoveries (sneering at "the precious secret of making water out of water") with a reminder of his worst deeds as a tax farmer. Marat showed no love for Lavoisier's "little disciple Fourcroy," whom he lumped among the "parasites" propagating the new doctrine.[24]

Marat's antielite invective gained him a cultlike status of devotion among the most radical elements of the Revolution. He easily won the election to the National Convention, with Fourcroy coming in second. That made Fourcroy Marat's *suppléant* for that district, destined to replace him should anything happen. And, of course, something did happen. Marat's assassination has emerged as one of the most iconic moments of the Revolution, immortalized in Jacques-Louis David's painting, as well as in literature, film, theater, opera, even video games. Every depiction contains the same core elements of the bathtub and turban, both efforts to treat a skin disease he had picked up in 1790 while hiding in the sewers. He kept his turban soaked in medicinal vinegar and his bath steeped in mineral and botanical medicines.[25] He conducted much of his business from his bath, and this is where Charlotte Corday, a supporter of the centrist Girondin faction, found him on 13 July 1793. She had hoped that killing the most radical revolutionary would pull the Revolution toward moderation, but instead it toppled into an even more desperate, radical phase.

Fourcroy entered a National Convention where everything was spiraling out of control. France waged war on all its borders, its troops immobilized by a lack of gunpowder. Its citizens rioted in the streets, complaining of the lack of basic necessities. But Fourcroy saw in these challenges "a welcome opportunity to develop all the strength of the [useful] arts," and particularly the new chemistry he developed with Lavoisier.[26] He and his former assistant Vauquelin wrote a series of technical manuals, distributed to the citizens of France, that gave detailed instruction on what they could do to help.[27] The topics of these manuals give an interesting indication of their priorities. Of the nine pamphlets they prepared, five concerned armaments or gunpowder, three discussed soap or the production of soda ash, and the final one promoted the overhaul of printed and written paper.

The soap question fell primarily to a nine-member group within the National Convention called the Committee of Public Safety, which was in charge of provisioning. They sent out questionnaires to every department of France, asking if they had any soap, if they had any manufacturers who made it, and, if so, what techniques and ingredients they used.[28] The ensuing report succeeded mostly in highlighting France's dependence on Marseille and Provence for their soap. Those regions, moreover, were not making enough soap for themselves and were loathe to export it. (Rouen, runner-up in soap production, used fish and whale oil, for a soap with a decidedly unpleasant smell).[29] The Committee of Public Safety also called upon citizens to submit suggestions for making soap without any soda ash or potash.[30] A group of chemists and apothecaries gamely ran experiments on the submissions—one recipe used potatoes, another a particular kind of white clay found in Provence. They also looked into making soda artificially from sea water. But nothing worked. The report made clear that there were no techniques that bypassed the use of soda ash or potash.[31]

Meanwhile, Fourcroy put together a Revolutionary Course on Gunpowder, cramming over 1,100 men into his amphitheater at the Jardin des Plantes.[32] He sent Vauquelin to search throughout France for the materials needed: saltpeter and alkali. He had enormous success with saltpeter, as citizens rallied to scrape the crumbly white earth off the corners of their cellars and damp brick walls, under piles of debris, and from the bottom of manure piles.[33] There were soon mountains of saltpeter waiting to be refined in Paris. But they had no such luck with alkalis, which were also in demand for soap. The Committee of Public Safety had put out a call for all patriotic citizens to give up the ashes from their fireplaces and to bring their buckets of used wash water to the saltpeter workshops, but this directive was largely ignored. As one saltpeter inspector put it, "Ashes are in some sense the property of women citizens, not all of whom are Republicans. For most of them that stuff is more precious than liberty."[34]

Gunpowder and soap remained locked in their rivalry, with the efforts of the chemists falling short and alkali remaining scarce. As the Committee of Public Safety continued to struggle with provisioning problems, one member within it, Robespierre, used the issue to garner a more prominent role. He echoed the outrage over shortages in his speeches. To succeed, he said, the Revolution must provide citizens with the basic necessities of life. The shortages and high prices came from "the perfidious designs of the enemies of liberty, the enemies of the people."[35] The Committee of Public Safety enacted strict antihoarding laws, known as the General Maximum because shopkeepers could not charge above a set price. There had been a price maximum on grain since May, but in September, it was extended to include soap, soda ash, potash, wine, eau-de-vie, vinegar, and more.

Anyone caught hoarding these items was labeled an enemy of the Revolution. Robespierre set up a streamlined justice system, the Revolutionary Tribunal, that could prosecute the large numbers of counterrevolutionaries he thought were undermining France. It dispensed with the inefficiencies that usually slowed down trials: defense counsel, witnesses, cross-examination. For the guilty, there was a single sentence available: death. Terror had come to France and many, including Blaise Laugier, would have their turn before the Revolutionary Tribunal.

PARIS, RUE BOURG-L'ABBÉ, FRIMAIRE, YEAR II (DECEMBER 1793)

Blaise Laugier was one of the few perfumers who stuck out the Revolution. By this time, his fellow perfumer Fargeon had fled Paris. Bourgeois had sold off his inventory and left, as well.[36] But Laugier stayed, contributing to the revolutionary effort as best he could. He and his wife, Marie-Jeanne, now had seven children. Their two oldest sons, Jean and Louis, had joined the Revolutionary Army. Their third child, Madeleine, had married and moved outside of Paris to Saint-Denis. Their four youngest sons, Antoine-François, Alexis, Blaise Jr., and Auguste-Victor, had come in quick succession, and ranged in age from nine to four when the Revolution began. Blaise himself served on the civil committee of his section, a neighborhood-level elected group that usually attracted the most fervently democratic.[37] To get some additional help in the shop, they posted an advertisement in June 1793, "looking for someone to take care of the books and correspondence."[38]

Laugier had always catered to a broader clientele than the perfumers who focused strictly on the royal court, and he embraced the

FIGURE 12. The Fountain of Regeneration, built on the ruins of the Bastille. It was the centerpiece of the Festival of Unity, held 10 April 1793, when the deputies gathered to ritually drink the waters of regeneration from the statue's breasts.

Revolution. He even came up with a new perfume based on the theme of "regeneration"—the Revolution's claim that by restarting their country from nothing, the people of France would themselves be born anew, stronger and better than before.[39] There was a literal component to this: the despotic hoarding of resources had kept the population weak and sickly, and a more equitable distribution would allow citizens to rise to their full force. There was also a strong symbolic component, perhaps best exemplified in a festival Robespierre organized in August 1793.

In the summer of 1793, Robespierre had asked the painter David to organize a Festival of Unity to commemorate the birth of the Constitution.[40] The theme of the event was "regeneration," and David set it at the Place de la Bastille, atop the ruins where the prison had been torn apart stone by stone. Everything centered around an enormous Fountain of

Regeneration: a giant woman, styled as an Egyptian goddess enthroned between two lions on a tiered platform, with water shooting out of the breasts she clasped before her. The Festival's ceremony began at dawn on August 10. As the sun rose, a choir of young girls clad in white sang the cantata *Hymn to Nature*. The president of the National Convention then mounted the fountain's pedestal.[41] He greeted the crowd and praised the statue before them as the incarnation of Nature, blessing the Revolution with her regenerative waters. Filling an antique chalice with water from the fountain, he poured it on the ground where the Bastille had been, re-baptizing it in the name of Liberty. He filled it a second time and drank it himself. The crowd watched as eighty-six representatives, one from each department of France, in turn filled the chalice and drank, accompanied by drum rolls, fanfares, and cannon fire.[42] The citizens, like the soil of France, were born anew with the regenerating waters.

Laugier began selling his own water of regeneration, or Eau Régénératrice. Perhaps best translated as "rejuvenating" or "revitalizing" water, Laugier recommended it for a wide range of salutary effects. It dissipated headaches, fortified the memory, and treated the vapors, dizziness, and ringing of the ears. It helped with rheumatism, gout, and any diseases of the nerves and joints. It stimulated the appetite, aided digestion, purified bad breath, and, when applied to the skin, treated razor burn, pimples, and dry patches. He particularly recommended taking it as a tonic that would "conserve and revitalize the natural heat" and reestablish the nervous and muscular functions of the body for anyone weakened by the overuse of their faculties.[43]

The advertisement called it a "very volatile alcohol impregnated with an *esprit recteur* drawn from diverse aromatic and balsamic plants." Most of the ingredients—including bergamot, bigarade, Portugal orange, mint leaves, tarragon, ground cinnamon, and rose petals—came from his depot in Grasse and were blended through a series of six separate infusions and distillations. The end result had a clean, revitalizing scent, more spicy than floral, with a strong citrus zest. "None of the virtues of the plants were lost in the amalgam, and all produce their salutary effects, and in this product, none is contraindicated by that of its companions." You could take it in a variety of ways. You could drink it, either by itself in small doses or mixed together with wine. You could dilute it with water and gargle it, douse it on a handkerchief, sniff it from a vial, or apply it to your skin. Pouring it in your bath was "a specific regenerator of all the animal faculties."[44]

Laugier continued to receive supplies from Grasse through the Revolution, but he, like everyone else, was impacted by the shortage of imported goods. In normal times, the entire third floor of his building was devoted to soapmaking, and Laugier Père et Fils sold a range of solid, liquid, and powdered soaps. This production was curtailed by the soda ash shortage, but they offered other products to contribute to the new regime of salubrity, including a number of cleansing vinegars and an Eau Sanitaire, ou Anti-contagieuse, to combat filth and contagion.[45]

Despite all of his demonstrations of revolutionary zeal, Blaise Laugier had his turn before the Revolutionary Tribunal. The Committee of Public Safety ordered his arrest on 28 December 1793, shortly after the National Convention had turned over executive authority, making them the de facto sole power of France. There was nothing personal about the arrest. They rounded up a number of other prominent perfumers, such as Jean Artaud and Bolard, and the celebrated vinegar maker Antoine-Claude Maille, placing seals over their account books and inventory.

Laugier was ultimately able to convince the tribunal he was neither a hoarder nor a counterrevolutionary. He was released within the month and his property returned to him. He was able to continue his business all through the Terror, even sending goods abroad. A commission now examined these exports to ensure no one was sending scarce resources out of the country. Laugier's trade included a case of toilet waters sent to Basel and a shipment of pomades to Geneva, both deemed acceptable, as they were "luxury merchandise and take most of their value from their workmanship."[46] It was not luxury itself the tribunal was against, but the hoarding of it.

Lavoisier was also caught up in the net of the Revolutionary Tribunal. After leaving the Gunpowder Administration, he had tried to stay out of politics but had been lured back with the prospect of reforming France's standard of measurements. He had always hated the nonsensical complications of the old measuring conventions, based on arbitrary things like the length of the king's foot and often varying from town to town. He went so far as to make up a system of his own for his personal work, with everything more rationally based on units of ten.[47] The Revolution provided an opportunity to make these reforms universal. He returned to Paris to head the Commission on Weights and Measures, where he worked with Fourcroy in a makeshift laboratory at the Louvre to create the new standards for the meter and the kilogram out of platinum.

But Lavoisier could not outrun his past. Although the General Farm disbanded early in the Revolution (and Lavoisier had resigned even earlier), the National Convention revisited the issue on 24 November 1793, ordering the arrest of all former Farmers General on the charge that they had not submitted their accounts in the allotted time frame.[48] The former offices of the General Farm on rue de Grenelle Saint-Honoré became a prison, with iron bars on the windows and mats on the floor to sleep on. Lavoisier shared a small room with his father-in-law, spending the days balancing the ledger books for their defense.[49] He refused offers of help to escape, convinced that he only had to show that he had followed every rule with unvarying precision. But the rules themselves had changed with the Revolution, and he was in more danger than he knew.

PARIS, GERMINAL, YEAR II (APRIL 1794)

The Revolutionary Tribunal's efforts to scour the country for hoarders had done little to solve either the soap famine or the gunpowder crisis. Robespierre, growing desperate, put the Committee of Public Safety in direct control of the gunpowder question, calling for a "revolutionary mode of production" that would replenish their depleted resources. He appointed another chemist, Jean-Antoine Chaptal, to lead the effort, choosing a manufacturer from the south of France who had experience in soapmaking and the production of soda ash.

Chaptal had been one of the few Frenchmen who dared smuggle a few ounces of barilla seeds into France, braving the penalty of death for the crime. In 1793 he had secretly planted them on the French coast, and after a season of meticulous care he harvested about twenty pounds of the plant. The alkali it produced, he reported, rivaled the products of Spain.[50] His experiments ended when he went back the next year to find his crop pillaged, but he was satisfied with the proof that barilla could be grown in France. He also tried his hand at producing soda ash artificially, founding a chemical works outside of Montpellier that he hoped would let him avoid paying the "onerous tribute" on products imported from abroad.[51] The enterprise never took off commercially, as his manufactured soda ash was still more expensive than any imported from Spain, but with trade from Spain shut down, it became of renewed interest.

Robespierre called Chaptal to Paris in the spring of 1794. Chaptal's first impulse was to refuse.[52] It was not that he lacked revolutionary spirit. He acknowledged that France could use a good revolution, but he admitted to a friend that he wished it had happened twenty years earlier, when he would not be stuck in the middle of it. His strategy so far had been to keep a low profile in the south of France. He had spent the early days of the Revolution in Montpellier, siding with the moderate Girondins, and had only avoided being arrested with the rest of them by fleeing to the mountains of Cévennes. Lately, he had been working as a regional inspector for the gunpowder program, traveling around Provence and Languedoc forming saltpeter workshops.[53]

But refusing was not an option. His friend and fellow chemist Claude Berthollet, who worked closely with the Committee of Public Safety, wrote that they "will have none of your refusal. Come! . . . any further resistance will be interpreted badly."[54] Chaptal arrived on 1 April 1794. That very night he was whisked before the Committee of Public Safety, who let him in on the severity of the situation. The lack of gunpowder had ground the French forces to a halt. There were fourteen divisions who were supposed to enact a massive, coordinated attack but were instead at a standstill, with some even losing ground. Robespierre ended the meeting by making it clear that Chaptal had exactly one month to remedy the situation and allow the campaign to proceed.[55]

The threat was more than implied. In case Chaptal thought his chemical expertise made him indispensable, he had only to look at the example of Lavoisier, who was then sitting in a jail cell awaiting trial. He and his fellow defendants had submitted their accounts on 27 January 1794. Lavoisier had prepared his with his usual meticulousness, but it did not matter in the end. The Convention judged them as a group and, seizing upon certain discrepancies, convicted them all of embezzlement. By this point, there were only twenty-eight of the forty Farmers General still alive, and the accelerating executions of the Terror had made the process startling efficient. It took thirty-five minutes, on May 8 at the Place de la Révolution, to guillotine them all.

Fourcroy had watched the trial silently from his seat in the National Convention. Thousands of words have been spent wondering if he could have saved the life of his friend and mentor. The question hung like a black cloud over his life and was raised even in his eulogy.[56] Later historians have grown more forgiving. The story emerged that he did, at one point, barge into a meeting of the Committee of Public Safety to plead for his friend's life, only to be met with a stony silence from Robespierre and a whispered warning from Prieur de la Côte d'Or in the hall after the meeting to drop the issue.[57] But whether this story was a later invention or not, it seems evident that Fourcroy, like everyone in Robespierre's vicinity, had reason to be concerned for his own life.

With Lavoisier's severed head as a memento mori, Chaptal settled into work. He, like Lavoisier, was a wealthy man of only moderate revolutionary enthusiasm, but he also had a keen sense of self-preservation. With little room to fail, he performed something of a miracle; he found the alkali he needed in a resource that France had in abundance—the leftover dregs of wine production. The fermenting process always left a sludge in the bottom of the barrel, called the lees. Vintners usually just threw it away, but Chaptal pointed out that one could press, dry, and burn the lees to produce *cendres gravelées,* or clavellated ashes. He had discussed this in the chemistry classes he had taught at Montpellier, but with no other source of alkali, it took on a new importance.[58] After Chaptal's arrival, the Committee of Public Safety had Vauquelin write another set of instructions to the patriotic citizens of France, this time targeting those in the Loire and other wine-growing regions, demanding the lees of wine "for the defense of Liberty."[59]

With a source of alkali secured, the effort to make gunpowder took on a frenzied intensity. Chaptal showed up for work every morning between 5 and 6 a.m., and left late at night, spending every waking moment overseeing a sprawling, manic scene. The Committee of Public Safety, now in total control of gunpowder manufacture, had moved operations from Essonne, a small town twenty miles outside of the city, right into Paris itself. They designated the former abbey of Saint-Germain-des-Près as the Workshop of Unity, dedicated to refining saltpeter, which then went on to the "gigantic" gunpowder mill at Grenelle near the Seine.[60] The "revolutionary mode of production" wasted little time on things like safety precautions. As Chaptal described his own operation, "everything was chaos, no more order, insufficient surveillance, inevitable accidents."[61] Workers rushed back and forth with gunpowder piled high in open wheelbarrows, the nails of their boots risking a fatal spark with each scuff on the hard cobblestones while nearby workers lit their pipes for a smoke.

When disaster struck, Chaptal was less surprised that it happened than that he had survived it. He had never been late, not a single morning for the first four months of the job, until 31 August, when he allowed himself a bit of extra sleep after coming in late the night before. He, with the rest of Paris, awoke at 6:45 a.m. to a deafening explosion that overturned all the furniture in his room. Grenelle had exploded, its many tons of stored gunpowder going up at once.[62] Arriving less than an hour later, Chaptal was stupefied by the scene. Over 500 workers and sixty horses had died in the blast, their incineration so complete that, sifting through the ashes, Chaptal reported scarcely finding two or three identifiable body parts.[63] The effects stretched for miles. Almost all the buildings in Paris had their doors or windows broken.[64] Houses collapsed in neighboring villages, and the sound of the blast reached Fontainebleau, thirty-five miles away.

Chaptal credited his narrow escape to "a kind of miracle," and his good luck did not end there. Robespierre had made it clear Chaptal's head was

FIGURE 13. The explosion of the Grenelle gunpowder works. Over 1,000 people died in what remains one of the worst industrial accidents in France's history.

on the line if anything went wrong, and he fully expected the worst. But Chaptal outlasted Robespierre. More moderate members of the National Convention, grown sick of the escalating purges, had banded together to force his arrest and execution. The day after the explosion at Grenelle, Fourcroy filled Robespierre's empty seat on the Committee of Public Safety, and Chaptal was safe.

The Terror was over, with France emerging from the state of all-out crisis. Soap returned to the shelves, and the riots ended. There was enough gunpowder to supply the French forces, and by the summer of 1794 they began a remarkable string of victories, pushing back their enemies on all fronts. A relative calm settled over the gunpowder workshops, and Chaptal even found time for some research of his own, presenting a paper to his colleagues on soap and the best methods for washing wool that has gone down in history as the first scientific work to use the Revolution's new metric units.[65]

4

The Miracle Waters of Cologne

MONTPELLIER, BRUMAIRE, YEAR V (OCTOBER 1796)

Fourcroy and Chaptal both emerged from the Revolution as heroes, demonstrating the power of chemistry to contribute to the glory of France. But what, exactly, would this new chemistry look like? The two men had different ideas. Fourcroy picked up the mantle of Lavoisier, arguing that all substances could be described by a chemical formula that counted up its constituent elements. Chaptal had been one of the earliest converts to Lavoisier's new chemistry, one of the very first to speak approvingly of it. But he was never one of the inner circle and did not share their conviction that all processes, both organic and inorganic, could be explained by chemistry. An unapologetic vitalist, Chaptal called Lavoisier's law of chemical combinations "the law of dead bodies" and insisted it applied only to the mineral realm. Living things, on the other hand, needed to account for their vital nature.

As France reestablished its medical schools after the fall of Robespierre, both Fourcroy and Chaptal had a chance to teach their distinct visions in a classroom. Guillotin had proposed three new Schools of Health, at Paris, Montpellier, and Strasbourg, to replace the abolished Faculties of Medicine. Fourcroy took the chair of chemistry at the School of Health in Paris, while Chaptal took the one in Montpellier. It was a full-circle moment for him. Years ago, at the age of eighteen, he had entered the Faculty of Medicine at Montpellier as a student. His uncle Claude was a prominent physician who taught there and had arranged a place for him.[1] Things did not work out as hoped. Chaptal had enjoyed his classes well enough, but he knew he would never be a doctor after confronting his first cadaver. He recognized the body—it was a young man he had played with as a boy, dead for four or five hours. On the first incision, he insisted throughout his life, the corpse put its right hand on its heart and shook its head. Chaptal dropped his scalpel and abandoned the study of anatomy for good.[2] He set off for Paris, where he mostly wrote poetry but also learned a little chemistry. In 1780, at the age of twenty-four, he returned to the south of France, where he taught chemistry classes and married a rich merchant's daughter. He used his wife's dowry and a generous gift from his uncle to start a factory making chemicals. Soda ash, as we've seen, was one of his first attempts, but there were many others, even more lucrative, before the Revolution interrupted everything.

Now, in 1796, he was back on the grounds of the former medical faculty, where the new School of Health asked him to address the students assembled for its inauguration. They must "throw the scalpel far away," he told them, if they wanted "to reach the sources of life."[3] Understanding a living thing meant studying its animation, how it lived in the world. To examine a body in isolation, as meat and bones, was to "know only the cadaver."[4]

Chaptal applied the same principles to chemistry. Chemists could not study the matter of a living thing after it was dead, he said, because they would miss the vital principles that so crucially distinguished it. Living things possessed an internal organization that was more important than their constituent elements. Chemists who "confined themselves to their laboratories" and "studied bodies in the state of death" had only, in his words, a "very incomplete knowledge."[5] Lavoisier had found success with his chemical laws, he acknowledged, but these laws applied only to the inert, mineral realm. In the animal and vegetable realms there was an interior force that could "subordinate to its wishes" the chemical laws.[6]

To study a plant, you had to catch it in the act of living, before the "volatile and invisible" dissipated. Chaptal identified three vaporous exhalations crucial to the process of life: oxygen, water vapor, and aroma or *spiritus rector* (he gave both the old name and Lavoisier's replacement).[7] Equally important were the fluids of life: blood in animals, and saps, resins, and essential oils in plants. Chaptal united these under the term *sucs,* which he identified as "the general vegetable humor, like the blood of an animal." While usually translated as "juices," it was in fact a broader category, including not only *jus* but all manner of organic fluids. Chaptal made clear that these vapors and fluids did not follow inert physical laws, for they moved upward through the plant against gravity. These were animated by a "vital force" that endowed organization upon the bodies.[8]

Chaptal tackled the great mystery of the organic realm that had stymied Lavoisier: fermentation. He was well aware of Lavoisier's claim that fermentation involved the conversion of sugar into alcohol, but he also knew that Lavoisier had never been able to get the reaction to occur using sugar alone. Chaptal thus looked to other properties of the grape, identifying four components that played a role: water, tartaric acid, sugar, and a ferment that he ultimately identified as the "vegeto-animal principle."[9] The sugar and the vegeto-animal principle developed together as the fruit grew and ripened, only becoming distinguishable in the process of fermentation, when the vegeto-animal principle acted as a ferment, or agent, of the process of turning sugar into alcohol.

Chaptal recorded his ideas in the first scientific treatise on winemaking, which among other things laid the groundwork for the modern concept of terroir. Wine, he said, could only be understood as the unique product of its region. Soil, climate, seasons, even the layout of the vineyard and its methods of cultivation all contributed to the formation of a product that could never be reproduced anywhere else.[10] But that did not mean science could not improve that process. On his suggestion, winemakers began adding more sugar to certain grapes to balance out the ratio with the vegeto-animal principle. Since the sugar was all converted to alcohol, the wine was no sweeter, but it was more robust, with a higher alcohol content. Known as chaptalization, the procedure is still widespread today.

In addition to treatises on wine, Chaptal wrote about vinegar and eau-de-vie, again repudiating Lavoisier's approach. Of Lavoisier's followers, Fourcroy had most thoroughly described the process of turning wine into vinegar, listing what he called the three necessary conditions: a sufficiently warm temperature, the presence of an acidic substance, and contact with air.[11] Chaptal added one more thing to the list: the vegeto-animal principle.[12] As in alcoholic fermentation, the transformation could not occur without this vital element. Chaptal performed numerous experiments in which he took old wines, stripped of their vegeto-animal principle, and left them open in the hot sun of southern France for hours at a time. They tasted sour and bad, to be sure, but not one of them turned into vinegar. He also studied the vinegar makers of Orléans, whose secret to success was to never fully empty their vats as they made the vinegar. They always left behind a sludgy glop known as the "mother of wine" on the bottom of the vat.[13] Chaptal proposed that this mother of wine was full of the vegeto-animal principle, and began publicizing it as the "Orleans process" or "French method." Both the quantity and quality of vinegar soared in France, and Chaptal promoted its use, suggesting it as a "healthy and agreeable drink." He recommended oil and vinegar to dress one's salad, a procedure novel enough that Charles Dickens referred to it as a "salade à la Chaptal."[14]

Chaptal placed vinegar among the "salutary remedies" that improved health and vitality.[15] These also included the products of distillation, and he provided recipes for a few of his favorite tinctures, liqueurs, and perfumes, such as an *eau divine* made with citron peel and orange-flower water, and a *rose crème* made with rose water and colored pink with cochineal. Aroma, or the *spiritus rector*, presided over these specific vitalities. In these affairs, Chaptal said "the nose is the best chemist that can be consulted."[16]

PARIS, 18 BRUMAIRE, YEAR VIII (9 NOVEMBER 1799)

Chaptal's family had been alarmed when he first told them that he wanted to drop out of medical school and travel to Paris, and it had only been the persuasiveness of his childhood friend, Jean-Jacques

Régis de Cambacérès, that allowed him do it.[17] The two spent a few glorious years in Paris and eventually formed a revolutionary club together in Montpellier. Like Chaptal, Cambacérès survived the Revolution through a combination of luck and self-reinvention. In and out of the revolutionary administration, he ended the 1790s back in Paris. He was a lawyer of exceptional charisma, a central figure in the Parisian gay scene, and known for the finest dinner parties in town.[18] The one he threw on 8 November 1799 was hardly the most lavish, but it was likely the most significant of his career. His guest that evening was Napoleon Bonaparte, a superstar young general of the revolutionary wars, just back from fighting in Egypt. He had been in Paris only twenty-two days, but a group of political insiders had already pulled him into a plot to topple the government. They hoped that Napoleon would lend both his military muscle and his immense popularity to the cause and then step aside. Cambacérès was on the outer circle of the original conspirators and knew only the vaguest details of the plan. But Napoleon was warming to him as an ally in his own coup-within-a-coup, ultimately grabbing power for himself, with Cambacérès as his righthand man. They dined that night in fine spirits, singing one of Napoleon's favorite revolutionary songs.[19] The next day, the eighteenth of Brumaire, they overthrew the government.

Napoleon named himself First Consul, after the highest office of the Roman Republic, and designated Cambacérès as Second Consul the following month. There was a Third Consul as well, an easygoing moderate named Charles-François Lebrun. But although Napoleon presented himself as the inheritor of the Revolution's democratic ideals, he rewrote the constitution to give himself the powers of a king. He brought back the tradition of the King's Council, renaming it the Council of State. Its twenty-nine members served as advisors, drafting legislation in the absence of the elected body he had destroyed. The work was all-consuming. The council met in various groups during the day and with Napoleon in the evening, in sessions stretching to four or five in the morning.[20] Napoleon appointed Chaptal and Fourcroy to the council, where they jostled for power in a tense rivalry.

Chaptal became a favorite of Napoleon. The First Consul sought him out at parties to chat about science and politics.[21] The admiration was mutual, for Chaptal greeted the new regime with high hopes, noting that "force succeeds weakness, order replaces anarchy." Within ten months, he was minister of the interior, replacing Napoleon's younger brother Lucien after a falling-out. This put him in charge of an immense and sprawling list of duties: public instruction, commerce, agriculture, manufacturing, museums, public works, hospitals, prisons, and more.[22] Nearly every institution needed rebuilding. As Chaptal put it, "after 10 years of anarchy which threatened to devour France, there existed almost no more social organization."

Education in particular needed an overhaul, as the abolition of guilds and apprenticeships had thrown a wrench in traditional paths of instruction. Chaptal put forward a comprehensive plan, which he submitted to the Council of State for approval.[23] The plan was not approved, undermined by Chaptal's rivals. Fourcroy had formed what another council member called a "standing cabal" against Chaptal with another member, Pierre-Louis Roederer.[24] Roederer's position within the council, director of Education, technically made him a subordinate of the minister of the interior, but he refused to even speak to Chaptal. Whenever Napoleon asked Chaptal about education, he replied, "Ask Citizen Roederer; he gives me no information." Tiring of his advisors behaving like children, Napoleon kicked Roederer off the council. This, in turn, made Fourcroy the director of education.

It was thus Fourcroy who established a new educational system in France, creating the structure of universities and *lycées* whose broad outlines are still in place today.[25] Fourcroy's chief obsession was the teaching of pharmacy. He had grown up in a family of apothecaries and worried over the fate of the profession without guilds. In the early stages of the Revolution, apothecaries had been included with physicians and surgeons in the new class of *officiers de santé* concerned with ensuring public hygiene. But when Guillotin began setting up his new system of medical schools for credentialing doctors, he explicitly excluded apothecaries from them. The field remained unregulated, with no standards or barriers to entry. It led to "the most complete anarchy," as Fourcroy complained before the council, leaving men who had spent years studying their art indistinguishable from "the most shameful charlatanism."[26]

Fourcroy's proposed solution, known as the law of 21 Germinal, Year XI, after the revolutionary day of its adoption, has been hailed as nothing less than the founding moment of the profession of pharmacy.[27] Reining in the empirics and charlatans, Fourcroy established strict guidelines for who could use the title of pharmacist. The first and most prestigious path was to attend one of the Schools of Pharmacy that Fourcroy established in Paris, Montpellier, and Strasbourg. After three years of coursework and three years working at an *officine,* or drug dispensary, one took an examination that tested both the theoretical and practical aspects of making pharmaceutical preparations. A second path, which gave a more restricted license limited to one's own department, involved spending eight years working in an *officine* and then taking the examination.

For Fourcroy, the field of pharmacy was rooted in the principles of Lavoisier's chemistry, and he filled the new School of Pharmacy in Paris with a close-knit group of like-minded adherents. He proposed his protégé Nicolas-Louis Vauquelin as the school's director, his cousin Jean-Pierre-René Chéradame as its treasurer, and two more relatives, Antoine-Louis Brongniart and André Laugier, as the chairs of Pharmacy and Natural History of Medications, respectively. They had all been apothecaries before the Revolution, most of them working together at Chéradame's apothecary shop on the rue Saint-Denis. Since then, their lives had only grown more entangled through a complicated sequence of intermarriage and a shared commitment to the new chemistry.[28]

I should point out that André Laugier, whose family came from the north of France, was not related to Blaise Laugier, who hailed from the south. Their lives had crossed in near proximity, as Blaise himself had been listed on the roll of apothecaries of Paris in the 1770s, working right around the corner from Chéradame's shop. But now their lives began to diverge again. The line between apothecaries and perfumers had been porous before the Revolution. Products such as Queen of Hungary water were both perfumes and medicines—it was senseless to even divide the two. But Fourcroy was now determined to draw a sharp distinction between them. Pharmacists, like doctors, would have the right to make medical claims, but perfumers, distillers, and liquorists would not.

Fourcroy's reforms gave pharmacists a monopoly on the sale of medicines and composite drugs. He also ordered the creation of a national pharmacopeia, the Codex, which would list the only permissible medical preparations, and anyone caught advertising their own personal concoctions would be punished. The target was what Fourcroy called "the crowd of secret remedies, always so dangerous."[29] He had in mind people such as the illiterate clockmaker known as Dol, who moved to Paris from Aix and began selling vials of some murky liquid he termed a "universal medicine," all while displaying a master's diploma, either stolen or forged, from Montpellier.[30] But Fourcroy's Codex also ran counter to the long tradition of guilds and monasteries passing down their own, unpublished recipes.

Fourcroy's efforts to outlaw secret remedies went virtually unenforced, as the market for them was too vast and demand was too high. Popular remedies constituted an important component of many people's health care. The legitimacy of such remedies varied, with a relatively small number of out-and-out frauds. Many of those selling elixirs possessed some kind of medical credentials, or had served as an *officier de santé* during the Revolution.[31] Others, like Blaise Laugier, inhabited a gray area where a lack of formal credentials did not necessarily imply a lack of relevant expertise. He was, after all, one of the best distillers in the city, using the highest quality ingredients. There was nothing fake about his Eau Régénératrice, and for someone seeking to revive their animal spirits it remained an attractive option. After a public outcry, the administration rolled back the law, adding a provision that the law did not cover preparations or remedies that had been permitted before the law's publication. But even this was not enough to curb the sale of popular remedies. And it did not help that the nation's leader was a profligate consumer of perfume, and convinced of its healing powers.

PARIS, TUILERIES PALACE, 1803

Napoleon was soon going through sixty bottles of perfume a month. He doused himself in it, bathed in it, carried a handkerchief scented with it, and even splashed some on the face of a man having a fit in front of him.[32] He drank it, diluted with water or wine, and kept a bottle beside him on the eve of every battle.[33] It was, he insisted, a necessary source of health and vitality. He cultivated an image of superhuman energy—sleeping less than anyone else around him and working longer—and perfume was what he used to keep himself awake and alert. His favorite method was to pour some into a hot bath, using the steam to envelop himself in the volatile aromatics. "One hour in the bath is worth four hours of sleep to me," he opined.[34] His attendants kept bathwater constantly hot, for he might decide on one at any hour of the day.

Chaptal grew accustomed to conducting business with Napoleon in the bath. The two of them toured France together several times, and every time they stopped, Napoleon's attendants would prepare the tub. He would hold meetings from it, or have an aide read him his mail, sometimes struggling to see through the billowing steam. It struck Chaptal as odd at first, but he got used to it. "This was his usage," Chaptal pointed out: "he claimed, I've said it elsewhere, that the water gave him back the forces that he lost through fatigue."[35]

As a young man, Napoleon had been rather indifferent to fashion. He had kept his hair long and powdered through the Italian campaign ("ill-powdered" in the opinion of the duchesse d'Abrante, who also reported that he "railed so loudly" against the young men of fashion).[36] He first cut it short while in Egypt, and even shorter after he became First Consul, when he began to wear it "à la Titus," like the founder of the Roman Republic. He had no patience for pomades, face-whiteners, or rouges. But he did have one thing in common with Louis XIV: he hated bad odors and did what he could to chase them from his presence. He generally had his windows opened first thing in the morning to get out the stuffy smells, and treated the room with vinegar and incense.[37]

The morning toilette had changed entirely in form, but remained a ritual of high regimentation. For Napoleon, it started with the bath, scented and very hot. He could remain in the bath for one or two hours, constantly adding hot water to replenish the steam. After the bath, he shaved himself, using a *savonnette* scented with orange or fine herbs.[38] Then he washed his hands with almond paste and rose soap, and his face with superfine sponges.[39] He cleaned his teeth with a wooden toothpick, brushed with powdered coral, then rinsed his mouth with a mixture of eau-de-vie and water.[40] Next came the most uniquely personal part of Napoleon's regime: the frictions, a habit that his valet claimed he picked up from the daily ablutions in Egypt.[41] He stripped naked, poured a bottle of eau de cologne over his head, and proceeded to rub his chest and arms with a rough brush.

FIGURE 14. Napoleon in the bath. In this satirical print from 1814, Napoleon is infusing the bath with the blood and tears of the people of France, who are being offered flowers of peace by a cherub.

He then passed the brush to his valet to rub his back and shoulders, yelling "stronger, like an ass!" if the brushing was not energetic enough.[42]

Napoleon's invoices listed perfumes such as Spanish jasmine and agarwood, and Windsor soap from England, but his overwhelming favorite was the fresh, light scent known as eau de cologne.[43] It owed its name to the city of Cologne, where it had first appeared over a hundred years earlier, the product of Giovanni Paolo Feminis, who had moved to German-speaking Cologne from a small Piedmont village sometime in the seventeenth century. He ran a prosperous shop importing fruit from his native Italy, whose sunnier climate offered delights like figs, grapes, and lemons. As a side venture, he began distilling some of them together in a recipe that capitalized on the many varieties of Italy's celebrated citrus fruit: bergamot, Italian lime, grapefruit, orange, lemon, and citron. He added some neroli and petitgrain, derived respectively from the blossoms and leaves of the bitter orange tree, which gave a fresh greenness. He finished with some rosemary, which softened the bite a bit, but the overall effect was still crisp and sharp. All of the scents were highly volatile and ephemeral, and did not last long after application. But this also meant there was not the slightest hint of overripeness or decay.

Feminis called the concoction *aqua mirabilis,* Latin for "miracle water," and sold it as a medicinal elixir, even submitting it to the Faculty of Medicine at Cologne to get its endorsement. This was a standard practice of the time. Many seventeenth-century apothecaries sold their own "miracle waters," and their recipes were carefully guarded secrets. In the eighteenth century Feminis sold his method to another Italian who had moved to Cologne from Piedmont, Giovanni Maria Farina. Farina came up with his own origin story, writing to his brother that he had found a fragrance that reminded him of an Italian spring morning, of "mountain daffodils and orange blossoms after the rain." Cologne had a strict system of guilds, from which Farina and his brother, as foreigners, were excluded. But as Catholics they were allowed to conduct business, and free of guild restrictions they ran a wide-ranging import-export operation advertising the selling of "Frantz Krahm," or French wares: silks, laces, stockings, wigs, scented powders, and toilet waters such as Queen of Hungary water.[44]

Farina's version of Feminis's miracle water remained a largely local phenomenon until the 1730s, when French soldiers passed through Cologne after fighting in Poland and picked up a few samples, which they brought back to Paris. Farina began focusing on exporting to France, adopting the name, Eau de Cologne, that his French customers had given it. Farina renamed himself as well. Having switched from Giovanni to Johann when he first moved to Cologne, he changed it again, to Jean-Marie.[45]

Business really took off after 1796, when Cologne was annexed into France. It had existed as a free, independent city for centuries, but once the French Revolutionary Army started winning battles, they pushed the boundary of France all the way to the Rhine. France abolished any restrictive commerce laws between them, and demand for Farina's Eau de Cologne soared. After Jean-Marie Farina's death, scores of different pretenders had stepped in to claim that they possessed the secret formula. Records show over 114 people claiming to be the Farina who sold "genuine eau de cologne," most having simply bought the use of the name. One man, Carlo Francesco Farina, sold the use of his name to over thirty different people, despite the fact that he had no relation whatsoever to the family of cologne-makers.[46] The original business, now run by Jean-Marie Farina III, tried to identify itself by its physical location, but this was not easy as there were no street addresses in Cologne. He wound up calling it *Johann Maria Farina gegenüber dem Jülichs-Platz* (which means "across from Jülich Square"). The name was usually shortened to *Farina genenüber* (or "Farina across from"), which had the unfortunate consequence of encouraging many of the other Farinas to name themselves after whatever was across the street from them. The French eventually assigned everyone specific addresses, not as street numbers, but as a long list of continuous numbers for every residence of Cologne. The perfumer who lived at number 4711, Wilhelm Mülhen, had paid to use the name "Farina," only to find out its provenance was fake. He continued producing eau de cologne under the name "4711" anyway, selling well in the crowded field.

Napoleon further invited the city of Cologne into France's fold, granting all its inhabitants French citizenship in 1801. He had begun to think of himself less as a modern-day Roman consul ruling over a republic and more as a modern-day Charlemagne presiding over an empire that stretched well into Germanic lands. He even made a replica of Charlemagne's crown and scepter, which he planned to use in a ceremony crowning himself emperor in December 1804. Before the event, he arranged to take a tour of the newly annexed areas along the Rhine, together with his wife, Josephine. Cologne scrambled to welcome them, building an eighty-foot-high pyramid and an obelisk in the marketplace, in honor of Napoleon's victories in Egypt. They decorated the city with garlands, trophies, and monuments, and honored him with allegorical paintings and long speeches praising him for lifting the city out of "darkness."[47] Napoleon, for his part, was only interested in visiting two places in Cologne: the famous Gothic cathedral in the center of town and Jean-Marie Farina's shop, across from Jülich Square.[48] He and everyone in his entourage bought large quantities of Eau de Cologne there, Napoleon's valet reported, packing much of it up to take back to Paris.[49]

PARIS, RUE BOURG-L'ABBÉ, 1808

The shop on rue Bourg-l'Abbé had been one of Jean-Marie Farina III's Paris contacts for years, importing the original Eau de Cologne to sell.[50] The store now went by the name Laugier Père et Fils, with father and sons presiding together over a thriving business. They had doubled the number of workers they employed and expanded their locations. Blaise's wife, Marie-Jeanne, had passed away in 1800, but his oldest son, Jean, had married, and he and his wife, Catherine, now had an infant boy, Édouard, born in 1807. The second son, Louis, now ran their factory in Grasse, which produced the pomades sold in the Paris shops.[51] They had also opened a second shop in Paris, on 290 rue d'Aboukir (recently renamed after one of Napoleon's battles), another middle-class shopping area less than ten minutes away.[52]

Paris was experiencing a "new boom in perfumery," according to an 1803 guide on style and fashion, which singled out Laugier Père et Fils as exemplifying the new trends. Laugier's store offered "all that the imagination can desire" when it came to one's beauty regime, while departing from the tradition of perfumers such as Fargeon, who catered to the aristocratic taste for scented leathers, gloves, fans, and many other things not directly related to perfume.[53] Laugier Père et Fils was now one of the two most popular perfume houses of the city, sharing with Houbigant the distinction of being the only perfumers on Paris's list of 550 of its most successful businesses.[54] Houbigant had, like Laugier, reoriented toward importing the Farinas' fashionable Eau de Cologne, with receipts for a three-month period in 1806 showing them supplying Napoleon with 162 bottles of the stuff.[55]

It was a shock for both when a stranger arrived in Paris in 1808 claiming to be the sole legitimate purveyor of Eau de Cologne in Paris. He called himself Jean-Marie-Joseph Farina, and claimed to be a great-grandnephew of the original Jean-Marie, part of a lineage that had stayed in the ancestral homeland of Piedmont. He had moved from Piedmont to Cologne by the period of French occupation, joining the crowded field of dozens of Farina pretenders. Once Napoleon had extended French citizenship to Cologne's citizens, Farina left for Paris. He married the daughter of a Parisian distiller, Pierre Claude Durochereau, and the two went into business together.[56] Durochereau supplied all the equipment and necessary materials, and Farina contributed the recipe and his famous name.

Farina and Durochereau opened two stores in Paris: one on rue Saint-Honoré next to Houbigant, and another on the rue d'Aboukir, next to Laugier's recent expansion. They advertised aggressively, plastering Paris with posters touting their products. "Superior to all others," these announced, boasting that they had been recognized by the "Institut Chimique" (an institute that did not exist, and never had).[57] Their underhanded tricks did not go unanswered. Someone kept pulling down the posters at the shop on rue d'Aboukir and pasting notices in their place that the enterprise had moved to rue Richelieu, which sent people knocking on the door of Farina's private residence. Farina was furious. He took out a half-page advertisement in the newspaper denouncing the act, convinced that it was the devilry of his rival perfumers, "Composers of bad Eau" who could not compete with his superior product.[58] He vowed to stay in Paris and continue the fight.

The one weapon he possessed was his name. Thanks to France's new trademark law, it was a powerful one. The *marques de fabriques*, introduced in 1803 by Chaptal's administration, were intended to cut down on the fraud and counterfeiting that had flourished once the guilds were abolished.[59] Manufacturers could submit any distinctive identifying "marks" for their brand—names, labels, and so on—to the Tribunal de Commerce, and no one else would be allowed to use them. Jean Laugier had made his own trip to the Tribunal de Commerce in 1806, registering forty-two different trademarks "to distinguish the different merchandise of their business."[60] When Farina arrived in town, he registered all sorts of variations on the Farina name, including "Jean-Marie Farina" and "Johann-Maria Farina." Neither one was technically his name, but they gave him the trademark anyway, and he began to aggressively pursue anyone using it for fraud. The "real" Jean-Marie Farina III remarked with alarm from Cologne that the interloper hurt his business in France.[61]

Laugier Père et Fils soon introduced their own Eau de Paris to rival Eau de Cologne. It had a similar palette, combining the bright citrus aromas of bergamot, citron, and Portugal orange with the green, woodsy freshness of neroli and rosemary. They hoped it would occupy a similar niche, selling for both its aroma and reported health benefits. But just as they introduced it, France began cracking down again on the sale of secret remedies. In 1810, pharmacists closed the loophole that had grandfathered prerevolutionary recipes, and now no one could sell secret remedies or compound medicines if they did not reveal their ingredients. But Napoleon, who used such remedies daily, did leave open another possibility. He established a onetime Commission of Secret Remedies, which would evaluate all the recipes sent to them, determine if any of them were "truly useful," and allow them to continue to make medical claims.[62]

The Commission of Secret Remedies got to work on 1 January 1811, overwhelmed with the number of submissions they had received.[63] They were brutal in their efforts to get rid of traditional nostrums. The

Carthusian monks had submitted the recipe for Grand Chartreuse, the 142-proof elixir with over 130 botanical ingredients whose recipe they had passed down in secret for generations, but the commission rejected it. Laugier's Eau de Paris did not make the cut. In fact, the commission approved only a single perfumer's claims: Jean-Marie-Joseph Farina and his Eau de Cologne.

Farina immediately published an entire book touting the product's numerous medical benefits.[64] He urged people to wash their face with it to prevent problems ranging from scales to pustules. Gargling with diluted cologne kept the breath from becoming fetid. Rinsing the gums with a stronger dose kept them healthy, and, in the case of a rotten tooth, applying a cloth soaked in cologne to the spot often brought back, "as if by enchantment," a peacefulness and ability to sleep. Headaches were treated by rubbing it vigorously on the temples. A wide range of ailments, including rheumatism, pleurisy, lumbago, and any general pain or weakness of the extremities, were treated with "frictions," after first applying cologne to the hands, brush, or flannels used for rubbing. A scented bath was effective for sluggishness and despondency—Farina suggested up to three bottles for a full-body bath, and one bottle for a foot bath. Cologne could be applied directly to cuts and contusions. For excessive flatulence, one could dip a piece of bread in the cologne and leave it on one's stomach. To prevent contagious fevers and plague, he recommended rinsing one's mouth with it, placing a soaked cloth behind the ears, and rubbing it on the hands and face. Taken internally, either adding a few drops to sugar or wine, or else taking a teaspoon by mouth, it could cure anthrax, pustules, menstrual irregularities, and any diseases involving a loss of vigor.

Farina had sent bottles of his *eau* to various doctors, pharmacists, and other medical professionals, and published the testimonials they provided in return. He even wrangled a testimonial from the chemist Claude Berthollet, who pronounced Farina's product "very good." Farina also insisted on including Berthollet's admission that "chemistry does not give the means of perfectly recognizing the elements which compose it." Farina pressed the point, emphasizing that his *eau* possessed virtues that even the best chemists could not explain and that when it came to the products of plants and animals, chemistry was "still in its cradle."[65]

The house of Laugier struggled to keep up. The second store on rue d'Aboukir closed down. The Napoleonic Wars brought further problems, as England put up a blockade that kept them from fulfilling any of their foreign orders. A bad harvest served as a tipping point, and on 11 February 1811, they found themselves in crisis. They asked the government for a loan of 120,000 francs, begging the emperor "to save from Disaster a house which enjoys a good reputation."[66] Napoleon granted the loan, replying, "We have always heard of the house of these fabricants as a house commendable for its loyalty and the scope of their affairs."[67]

Blaise Laugier retired in 1812, leaving his sons to deal with the transformed landscape. He returned to Grasse, purchasing an enormous house on the main square. His oldest son, Jean, took over affairs in Paris, and soon became one of the first perfumers in France to take out patents on his products. Although the patent system, the *brevets d'invention,* was created in the Revolution, it was not until after the crackdown on secret

FIGURE 15. An advertisement for eau de cologne, stressing that it was a "German Invention" that achieved "French perfection." This particular product was from a later purveyor, Théodore Chauloux.

remedies that perfumers began to use it. On 15 September 1812 Jean Laugier applied for a patent for his own recipe of eau de cologne—the third patent ever granted for any kind of alcohol-based perfume.[68] The patent specified his scientific credentials, emphasizing that they worked "according to the procedures of Baumé, perfected by M Chaptal." The eau de cologne was just the first of several brevets that Laugier submitted. Others included Eau de Paris and their Eau Régénératrice.[69] The third son, Antoine-François, moved down the street to 26 rue Bourg-l'Abbé. He entered into a partnership in 1814 with a businessman, Guibert, for the fabrication of Eau de Cologne, by means of a brevet that he obtained in his name from Jean-Marie Farina III (a direct descendant of the original, in Cologne).[70]

Fourcroy's collaborator and former assistant, Vauquelin, also applied for a patent for his own eau de cologne. His had been one of the leading voices encouraging the separation of the perfume business from medicine, and his actions point to the paradoxes of his position. There was no question that certain plant and animal substances could have profound effects upon the human body. This was the entire basis of the field of pharmacy, and plainly obvious to anyone who had, say, nibbled on a belladonna plant and had intense hallucinations, or sampled some ipecac root and immediately started vomiting. Apothecaries had long called these the "active principles" of plants, named for what they called their "action upon the animal economy." But what, exactly, produced these effects, and how could the new chemistry begin to account for it?

5

The Problem of Vegetation

L'AIGLE, FRANCE, 1803

It was one o'clock in the afternoon when the inhabitants of L'Aigle, eighty miles north of Paris, heard a deep rumbling. It seemed almost like thunder, although the sky was clear and there was no sign of a storm. Instead, looking up to the sky, they saw an immense fireball streaking past. As it headed toward a nearby field, it suddenly blew into thousands of pieces, with a series of deafening explosions that were heard fifty miles away. The villagers ran to the fields to find them littered with stones, gouged deep into the earth as if by great force and hot to the touch. They could hear an audible hissing reverberate in the field and smell sulfur wafting from the stones. It was alarming and strange. Yet theirs was an enlightened age, and they wasted little time before alerting the scientists. Within two weeks, Fourcroy read their report on the floor of the Academy of Sciences.[1]

They were not the first people to see rocks falling from the sky. Meteorites were a well-known phenomenon, although the learned literature tried to distance itself from the idea that they came from the heavens. The Aristotelian system, with its strict division between the celestial and terrestrial, necessarily classified them as a sublunar phenomenon, probably caused by volcanoes or stormlike weather disturbances. Even after the Aristotelian system was dismantled, Enlightenment philosophers tended to dismiss tales of their heavenly origins as silly, peasantlike credulity. Lavoisier thought the new chemistry could show their terrestrial origin. He was able to detect pyrite in meteorites which he thought attracted lightning to them, earning them the nickname "thunder stones."[2] Newspapers echoed this stance in their reports, assuring their readers that "We are not at war with the moon," and "Everything happens in the air, within that immense laboratory where lightning, hail and storms gather."[3] But the issue was far from settled. Figures as lofty as Pierre-Simon de Laplace, France's preeminent astronomer, leaned toward the belief that meteorites did have an extraterrestrial origin.

The fireballs at L'Aigle deserved an investigation. Chaptal, minister of the interior, sent the Academy of Sciences' newest member, Jean-Baptiste Biot, who sat for his first session on the very day the meteorite landed. He had arrived at the academy through a rather improbable path. Eighteen when the Revolutionary Wars broke out in 1792, he had immediately joined the army and received officer training as part of the first class of students at the newly created École Polytechnique. After the war, he married and moved to his wife's hometown of Beauvais to teach high school mathematics. Yet he continued to nurse what he later called his "immoderate ambition to penetrate the highest regions of mathematics, where one discovers the laws of the heavens."[4] Not content to sit on the sidelines, Biot wrote to Laplace, asking if he could get an advance copy of his current work in progress. It was an audacious move. Laplace was a legendary, rather terrifying figure, hailed in France as the successor to Newton. The work in question was his magisterial *Celestial Mechanics*, which laid out the motion of heavenly bodies with harrowing mathematical sophistication and solved all the lingering instabilities that had plagued Newton. Laplace at first dismissed his request but eventually allowed Biot access to the page proofs as they came off the press, happy to know someone who knew enough math to catch any misprints. Laplace took him under his wing, arranging a teaching position for him in Paris at the Collège de France and supporting his candidacy at the Academy of Sciences.

Biot's trip to L'Aigle was his first scientific expedition, and he took it seriously. He spent nine days in the village, even missing the birth of his first child. He interviewed anyone he could find, taking pains to assemble a diverse group that had not coordinated their stories. In addition to a number of peasants, he talked to "enlightened men," "a very respectable gentlewoman, who has no interest to impress anybody," "two clergymen," and "an elderly lady."[5] The consistency and trustworthiness of their stories made what Biot called "moral proofs" for the claim that the rocks fell from the sky. There were also "physical proofs." He traversed the nearby fields to produce a detailed map of where the stones had fallen, counting over 3,000 of the mysterious projectiles. He compared them to the local rocks in the mineralogy collection kept by an engineer in the area and found nothing remotely similar. He also visited every foundry, factory, and mine in the area, but found that none of their products could be confused with the stones. He searched for signs of volcanic activity but found none. The facts pointed to a clear conclusion: these rocks were from outer space. His report, published and distributed widely by the Academy of Sciences, became the crucial document establishing the extraterrestrial nature of meteorites.[6]

Meanwhile, Fourcroy and Vauquelin had completed their own chemical analysis of meteorites.[7] They had obtained samples from all over the world and were struck by the similarity of their mineral composition, and particularly by the presence of the element nickel, which was virtually unheard of in terrestrial stones. Here was another strong piece of evidence for the extraterrestrial origin of meteorites. Such evidence was a triumph as well for modern chemistry and its analytic technique of breaking substances into their constituent parts. So far, however, all of the victories had been notched in what was called, following Linnaeus, the mineral kingdom. The other two kingdoms, vegetable and animal, remained unconquered.

Would the new chemistry ever be able to explain living things? Lavoisier had certainly tried. He had logged hour after hour combusting a wide variety of plant matter in sealed chambers to see what the products would be. But he struggled to say much more than that they all contained carbon. His new rules of nomenclature, so useful and precise for the inorganic world, failed him in the domain of living things, which continued to use a traditional vocabulary of volatile and essential oils, resins, balms, gums, and extracts. Fourcroy, and the circle of ex-apothecaries around him, took up the problem as their particular challenge.

Vauquelin laid out a manifesto for what he called the "problem of vegetation." It was, he said, a "beautiful problem," "a true chemical problem." All you had to do was reduce a plant to its primitive principles: hydrogen, carbon, and oxygen. These principles existed in varying proportions in a state of equilibrium, which accounted for each plant's particular odor, flavor, color, and so on. Fermentation, to take one example, was for Vauquelin a purely chemical process in which a "change in equilibrium" gave rise to new combinations between the principles.[8]

Fourcroy and Vauquelin did much of their work together at this point. They studied the pollen of date trees, the juices of onions, the sap of banana trees.[9] They spent six years investigating the germination and fermentation of seeds.[10] They prepared an extract of ants, infusing them in alcohol and redistilling the product. They determined (incorrectly), based upon the smell, that the product was acetic acid, or vinegar.[11] They studied bile, snot, and tears, gallstones and bezoars, and spent particular time on the urine of newborn babies. They studied the brains of cows, sheep, and humans, dissolving the grey matter in alcohol to see what it was made of.[12] They studied the process of growth in living bodies and its opposite, decomposition. Fourcroy had been in charge of moving the corpses from the Holy Innocents Cemetery to the Catacombs, and he found a great number of these bodies had transformed into a white substance, fatty and combustible, seemingly like spermaceti found in the heads of sperm whales. With more work, he found that this metamorphosis takes place for all animal matter preserved from contact with air in humid places.

One plant of particular interest was the bitter almond, which combined an alluring scent with a deadly potency. The bitter variety of the almond tree was more common in the wild than its sweet cousin, and its fruit had a deeper, more intense smell that was highly prized. Yet a small handful of these almonds, as few as ten, could kill a person. Ancient sources warned of their deadliness, yet also instructed how to use their scent in unguents and balms. One Egyptian papyrus from 1600 BCE contained a "Recipe for making an old man into a youth," which detailed how to make a poultice from crushed bitter almonds that would rejuvenate the skin and make wrinkles and age spots disappear.[13] Greek, Roman, and Arab recipes used them as well, and over 3,000 years later at the court of Versailles, the best-kept secret for keeping skin soft and youthful was the *pâte d'amandes,* a carefully prepared paste of ground bitter almonds.

But its deadly side was hard to ignore. Indeed, distilling bitter almonds produced one of the most potent poisons known in the eighteenth century, killing its victims in a matter of seconds. French chemists had named it prussic acid in the 1780s, after a Swedish pharmacist, Carl Wilhelm Scheele, found it in a preparation made from the dye called Prussian blue. The dye itself had no almonds in it, being made from a rather unpleasant combination of blood, potash, and iron. But the smell was unmistakable, and its action as a poison was identical, too. Scheele died soon after his announcement, but French chemists took notice and continued investigating.[14]

Fourcroy and Vauquelin were particularly excited about studying prussic acid, filling their shared laboratory with samples of plants that produced it when distilled: bitter almonds themselves, of course, but also peach pits, cherry stones, and apricot seeds. Vauquelin even went so far as to work backward, turning the prussic acid he had made from apricots into Prussian blue pigment to prove their relationship. But in the process he noticed something odd about the familiar smell: "One finds the effects are not proportional to the cause."[15] Prussic acid has a rather faint smell, and the samples he was getting from the almonds and fruit had a relatively low concentration of it. Yet the samples themselves smelled quite strongly. Perhaps, he noted, the smell of almonds did not come entirely from the prussic acid. He let the question drop and moved on to other things, but it stayed lodged in the mind of his assistant at the time, Pierre-Jean Robiquet.

Prussic acid was renamed "cyanure" from the Greek word for blue, which then became cyanide in English after the chemical revolution.[16]

It marked another triumph for the new chemistry, yet Robiquet remained bothered by the mystery of the "prussic odor." Just what was the relation between the deadliness of the substance and the smell that always seemed to accompany it? The question plunged him deep into the paradox of smell. Lavoisier and Fourcroy, he said, had dismissed aroma to the realm of "imaginary beings," nothing more than volatile compounds that quickly dispersed.[17] Yet for something that did not exist, smells seemed to possess an uncanny ability to affect people's bodies, or, as he put it, "their action on the animal economy."[18] The old concept of the *esprit recteur* had seemed to offer a more satisfying explanation, emphasizing that these subtle fluids had been capable of decidedly real effects.

PARIS, SAINTE-PELAGIE PRISON, 1814

Few plants had more desired effects upon the body than opium. It was not hard to find some in Paris if you wanted it. The stuff was perfectly legal, unregulated even. And while you were most likely to find it on the shelves of a pharmacy, there were no restrictions on who could sell it. Laugier and his sons sold it alongside other cures listed for "headaches," injuries, or other ailments. Indeed, its distinctively foul smell meant it was often sold as part of a tincture with more fragrant botanicals, a specialty of perfumers.

The use of opium stretches far back into human history, but its rise in Europe paralleled the use of distilled liquor. Paracelsus, the unflagging supporter of distilled spirits, was also central in securing its popularity. Calling it "the stone of immortality," he administered it widely to his patients and experimented with the substance on both himself and others.[19] The little black pills looked, according to an assistant, like "mouse excrement," but Paracelsus boasted that "with these pills he could wake up the dead."[20] Some hundred years later, the English apothecary Thomas Sydenham was equally enthusiastic, noting that of all the remedies known to him, "none is so universal and efficacious as opium."[21] Its unpleasant, bitter taste was a problem, however. To mask this, Sydenham began selling an elixir in which he compounded the opium in wine with saffron, cinnamon, and cloves: laudanum. Its popularity as a medicine soared, although, as Thomas de Quincy's *Confessions of an English Opium Eater* (1821) made clear, it could also induce dependency.

France loved opium, too, importing 2,000 pounds of it in 1803 and 3,000 pounds in 1807.[22] Apothecaries had long appreciated its ability to relieve pain, and sold it in various tinctures, pastes, and powders.[23] Antoine Baumé discussed it at length in his *Elements of Pharmacy*, explaining that the best product came from Persia and Turkey, in pungent bricks made from the paste extruded from the oversized seed of the poppy plant. He gave detailed instructions on how to purify the paste, distill it, and then let it "digest" in a bain-marie for weeks or even months to make an opium extract. He provided several different recipes for the extract, including not only Sydenham's laudanum but others, such as one from the Paris Faculty of Medicine using the "spirit of amber," one from the Abbé Rousseau using honey, and one from the physician Langelot using quince juice.[24] He also discussed another common way of preparing opium: mixing it with honey to make a paste or electuary that was typically applied to a toothache to ease the pain. These were called opiats, and the term eventually broadened to mean any kind of paste mixed with powdered medicaments applied to the teeth. Laugier sold these as well, in a corner of the shop devoted to dental hygiene.[25]

How to account for opium's effects? Molière had ridiculed the seventeenth-century doctor in his play *Le Malade imaginaire* who tried, with useless circularity, to attribute its sleep-inducing properties to a "soporific virtue." Baumé had focused on its "noxious and narcotic aroma," equating it with a "*principe recteur*" responsible for its effects. Fourcroy and Chaptal also paid particular attention to its bad smell. They hoped, based on Baumé's work, to be able to separate the "calming principle" of opium, so useful for medicinal purposes, from the "drunkenness and stupor" that caused all the problems.[26] They each took their own approach. Fourcroy tried to separate the substance into its various components: a resin, a salt, an essential oil, and so on.[27] Chaptal treated the white, milky

FIGURE 16. J. J. Grandeville's depiction of "Poppy," in his 1847 book *The Flowers Personified*. She is putting all the insects to sleep with her "powerful, and even dangerous" narcotic juices.

sap of the poppy seed as one of the *sucs*, or organic fluids, that contained the vital vegetable principles.[28] But neither was able to find the source of the calming effect.

A pharmacist on rue Saint Honoré, Jean-François Derosne, had a breakthrough, although not precisely the one he wanted.[29] He had also read the work of Baumé and followed his methods of preparation, discovering in his investigations a new kind of salt that no one had identified before.[30] Wanting to see what sort of effects it had, he fed some to several dogs. The doses were very small, four decigrams, but all the dogs were violently sick, as if they had been given large doses of opium.[31] They were vomiting, falling, and convulsing, and only improved when he forced vinegar down their throats. The new substance seemed to impart all of the stupor with none of the calm. On the plus side, Derosne noted, he could now recommend vinegar as an antidote to opium poisoning. But the "calming virtue" of opium remained elusive, and Derosne urged physicians to look for it in the substance left over after his new salt had been removed.

There was someone in France who had succeeded in isolating the "calming virtue" of opium, but he spent most of Napoleon's reign in prison. Not because of the drugs. Those, of course, were all perfectly legal. But for the far more serious crime of quarreling with Napoleon. Armand Séguin had been a chemistry student working in Lavoisier's lab at the Arsenal. He was sometimes part of the experiments themselves, as when he donned a rubber mask to help test the vitality of the air (see Figure 10). Once the Revolution started, his close friend Fourcroy had touted him as someone able to cure leather in a fraction of the standard time, and the Committee of Public Safety contracted him to provision the revolutionary army. He became the sole provider of leather to the largest army in the world, which would be continuously at war for over a decade. The money rolled in.

Séguin was not particularly discreet with his new wealth. He owned a large island in the middle of the Seine that he named after himself. He also began some eccentric collections: rare musical instruments left scattered about his mansion, expensive horses that ran wild on his estate. Napoleon had little patience with what looked to him like war profiteering. He ended the contract and, in 1804, presented Séguin with a bill to repay millions of francs in restitution.[32] Séguin refused, calling it political persecution. What followed was the biggest financial scandal of its day. In the end Séguin, one of the richest men in France, was sentenced to the debtor's prison of Sainte-Pelagie.

Séguin had given a presentation before the Academy of Sciences in 1804 describing how he had isolated the component of opium responsible for its calming effect, but the scandal broke before he published his results. He stayed in prison, refusing to pay. He used his own money to luxuriously refurbish the apartments where he was detained, and his constant influx of high-society guests made it one of the most fashionable salons in Paris.[33] His strategy was to outlast Napoleon, and while it took ten years, it eventually paid off. Finally defeated by the combined forces of Europe at Waterloo, Napoleon headed into exile in 1814. Séguin was soon out of prison.

He finally published "On Opium," in which he described how he isolated a precipitate of "white, prismatic crystals" from the opium he had purchased and concluded that it "could only be considered as a new, entirely particular vegeto-animal matter."[34] The problem was that by the time his essay appeared, someone else, in Germany, had found the same substance. What's more, they had given it a catchy name: morphium, after the Greek god of dreams.

Friedrich Sertürner was a twenty-year-old pharmacist's apprentice when he began working with opium.[35] He had isolated the same white crystals and proceeded to try them out on stray dogs and rats he caught in the basement, which all kept dying. His next move was to try it himself with three of his friends. They all took three doses of a tenth of a gram over a forty-five minute period, drinking the morphine dissolved in alcohol. After the first dose, he said, the faces of his companions turned red and "their vital forces seemed exalted."[36] After the second, their heads spun and they all felt like vomiting. After the third dose, Sertürner fell into a dreamlike trance. Noticing that the others were losing consciousness, he ended the experiment by making everyone drink vinegar and vomit.[37] He wrote a letter to the editor of a pharmacy journal in 1805, but was generally ignored.[38] He had no university training, was working on self-made equipment, and was experimenting almost entirely on himself. He complained later that "in Germany," he had been "basely rejected, yes, maliciously scoffed at and forgotten."[39]

After Séguin's publication in 1814, French chemists resurrected Sertürner's neglected work. Joseph-Louis Gay-Lussac published a translated version in the *Annales de chimie* in 1817, renaming the molecule "morphine" to better fit with the emerging nomenclature and flagging the enterprise as being of "the greatest importance."[40] Vauquelin threw himself into the study of opium.[41] He had worked for years on the question of what gave plants their special powers, and this was finally a significant break in the case.

There was something intriguing about the chemistry of morphine. For Lavoisier, the most fundamental divide in the new chemistry was between acids and bases, and products derived from the vegetable world had always only been acids. While it was true that potash and soda ash came from plants and were famously alkalis, or bases, most chemists, including Fourcroy, had attributed their alkaline nature to minerals drawn from the soil, rather than the vegetable material itself, which after all had to be burned to make the ashes.[42] But morphine was itself a base, something never before seen in the organic realm. Gay-Lussac suspected that it represented an entire class of molecules, one he called the organic salifiable bases. (*Salifiable* meant that it would react with an acid to form a salt.) Now called alkaloids, they opened a promising new path of research.

Vauquelin and his students plunged in. Fourcroy had died in 1809, but he and Vauquelin had together trained a new generation of pharmacists well schooled in Lavoisier's principles. Louis Jacques Thénard had been their most promising student, taking over Fourcroy's teaching positions at the École Polytechnique and the Faculté des Sciences, as well as Vauquelin's position at the Collège de France when he retired in 1804. Robiquet, who had assisted Vauquelin with his work on bitter almonds, now had a position of his own at the School of Pharmacy, as the chair of the Natural History of Medicines. He began the task of verifying all of Sertürner's claims, opening a new pathway for studying the active principles of plants.

Two more former students of the School of Pharmacy, Pierre-Joseph Pelletier and Joseph-Bienaimé Caventou, soon discovered another alkaloid, which they named "Vauqueline" after their mentor.[43] They had been working with the Saint Ignatius bean, chosen as one of "the most active plants of the materia medica."[44] The bean was long known for its fast-acting poison that stopped the heart, and it had been used in its native Philippines to poison the tips of blow darts. Pelletier and Caventou isolated its poisonous component in 1818. When they presented their work to the Academy of Sciences, the members pointed out that it was among the most fatal poisons known, and suggested that "a beloved name should not be applied to a harmful principle." Pelletier and Caventou came back with a new name, strychnine, which stuck.[45]

A cascade of new alkaloids appeared in the following years. Within two years, Pelletier and Caventou had found at least five more.[46] The most significant of these was quinine, which they identified as the active principle of the cinchona tree. The bark from this South American tree was one of the favorite medicines in the apothecary's cabinet. It was the most effective fever-reducer available, and the isolation of quinine, its "febrifuge principle," was a notable coup.

Robiquet added to the list of alkaloids. He continued work on opium and found it contained other alkaloids as well. He named one codeine, and rebranded Derosne's "peculiar substance" as the alkaloid narcotine (now called noscapine). German chemists also caught the fever for alkaloid discoveries. When the German poet Johann Wolfgang von Goethe asked his friend, the chemist Friedlieb Runge, to investigate his favorite drink, coffee, Runge found yet another alkaloid—caffeine.

It was as if the secrets of the apothecary cabinet had been unlocked. The oldest and most sought-after remedies revealed their active principles, one after another. It explained medicinal plants whose reputation for mind-altering powers stretched back to antiquity. Deadly nightshade, hemlock, mandrake root—all powered by alkaloids. It explained the high-demand drugs of the New World, whose allure had been strong enough to ignite a global trade and reorganize the world's population. Coffee, tobacco, cinchona—all were tamed and understood, their active principles revealed in caffeine, nicotine, and quinine. It was an exciting time to be a chemist.

6

A Temple of Industry

PARIS, RUE BOURG-L'ABBÉ, 1824

Rue Bourg-l'Abbé had lain in the shadows of the abbey of Saint-Martin-des-Champs for as long as it had existed. The abbey's monks had inhabited the spot for over a thousand years, in a complex of buildings that stood first outside the city of Paris and then within it as the city expanded. They had begun with a chapel, a refectory, and some simple living quarters, but in the twelfth century added an ornate Gothic church to rival the cathedral Paris had just built at Notre Dame. The order continued to grow in importance over the years, soon becoming one of the richest religious establishments in France. By 1789, it collected revenues from nearly a hundred other priories, vicarages, and parishes all over France, despite the fact that only twelve monks now lived on the site.[1] This blatant accumulation of wealth—a form of ecclesiastical hoarding as egregious as the aristocratic kind—caught the eye of revolutionaries, and in 1790, the National Assembly confiscated the properties. They planned to turn the abbey into a school for the industrial arts, named the National Conservatory of Arts and Crafts (where "arts and crafts" is perhaps best understood as "industry and trades"). With the church at its center serving as a museum for the latest technology and machinery, it would be a temple to a new industrial age.

The plan had not gone smoothly. The site went unused for most of the Revolution, serving briefly as a prison and then as an improvised weapons manufactory. Under Napoleon, it became a kind of warehouse of confiscated machinery, its rooms full of expensive devices that had once graced royal salons. There was a life-size mechanical automaton, originally a gift to Marie Antoinette, which would play eight of the queen's favorite songs on a dulcimer, run by machinery hidden in its skirts. There were "mechanical paintings" in thick gilt frames, animated by systems of gears and rods. One showed the collection of lavender, with a horse-drawn cart moving slowly through a field, flanked by cows, sheep, and goats all moving and voiced by automated bleats. The church finally opened as a museum in 1806, allowing visitors to file through on Sundays and Thursdays to view the collections. Alongside the baubles of the rich were more serious machines, such Jacques de Vaucanson's device for automating the weaving of silk, and Joseph Marie Jacquard's programmable loom, where one had only to feed the machine a series of punch cards and it could weave a cloth in any design imagined.[2]

It was not until the 1820s that the National Conservatory finally began offering classes on a regular basis. They were free and open to the public, and while their subjects—chemistry, mechanics, and industrial economy—rested on fundamental principles, the school stressed they were all "applied to the [industrial] arts" and intended for working men. Édouard Laugier, the only child of Blaise's oldest son, Jean, had just turned seventeen and was helping his parents make perfume. Did he ever attend lectures, or take a day off to visit the museum down the street? No one took attendance. But the classes were enormously popular, drawing people from all backgrounds. Sadi Carnot, for example, was an army officer on reserve duty in Paris when he witnessed a demonstration of the steam engine in the chemistry class in 1824. The device entranced him, with its extraordinary ability to replace the work of muscles with machines. The book he wrote that same year, *Reflections on the Motive Power of Fire*, is a foundational text in the science of thermodynamics. Through it he explored humanity's newfound power to trade heat for work and the new era such power promised, in which burning coal would replace human labor.

Despite the excitement generated by the steam engine, the chemistry instructor at the National Conservatory, Nicolas Clément, had another

FIGURE 17. Visitors view displays of machinery in the nave of the church of Saint-Martin-des-Champs. Although this engraving was made in 1880, the church had been serving as a museum of industrial machinery since the early nineteenth century.

subject close to his heart: distilling. Clément, an industrialist, ran a beet-sugar refinery and a distillery for making alcohol from potatoes. He spent particular time demonstrating a new distilling apparatus recently deposited at the conservatory by Charles Derosne, brother and assistant to the opium-experimenting Jean-François Derosne. It was an innovative design, whose patent Derosne had purchased from the inventor Jean-Baptiste Cellier-Blumenthal in 1818 for 1,200 francs.[3] Clément called it the Derosne still in his class, emphasizing the tall tower that let the vapors rise slowly in stages, gaining in concentration as they got to the top. Édouard Laugier, whether seeing the still in Clément's class or not, began working with similar designs.

The classes at the National Conservatory embodied the industrial ethos that now permeated Paris, and Laugier Père et Fils embraced it fully. Business was once again brisk. Jean ran the flagship store in Paris, with Alexis Louis at the factory in Grasse and Antoine-François down the street, selling eau de cologne. The fourth son, Victor, opened a store selling perfume at 21 rue d'Hanovre, in a rising fashionable district. The only daughter, Louise, lived nearby on the rue Saint-Denis, widowed after the death of her husband, Alexandre Theodore Faveau. Jean's son, Édouard, had become indispensable, spending much of his time developing new products in the laboratory in the back of the house.

The Laugiers took out a paid announcement in a leading paper, *Le Constitutionel*. These were the early days of advertising, when the copy read like a news report but was part of a separate insert distributed with the paper. It began with a praise of French industry, noting that "among the most agreeable and useful" of its products was perfume in general, and the offerings of Laugier Père et Fils in particular. "Not a year goes by that does not see some new invention come out of their laboratories: the last months of 1823 and the beginning of this year have seen several very interesting ones."[4] There was a new Huile à la Neige for the hair, to protect it from humidity and keep it in place, and a new Poudre de Laugier Père et Fils to clean the teeth and protect them from decay. But the most exciting was the new Eau Cosmétique, which they promoted as combining the intertwined goals of health and beauty.

This Eau Cosmétique was a product of the new stills, with a higher alcohol content than had previously been possible. It was, they said, "both more spirituous and more astringent" than previous eaux, perfect for slapping on the face after a shave. They particularly recommended it after a bath. Regular bathing was just beginning to catch on among the middle class, and Laugier's ad copy included what amounted to instructions for how to do it right. A range of negligent practices, they warned, risked death for the insufficiently cautious. These included bathing with a full stomach, when it was too humid, when one was covered in sweat, or when one's humors were agitated by violent passion. But done right, a bath could be a source of both pleasure and health. They noted several salutary effects, such as opening the pores, freeing up the paths of circulation, as well as "dividing thick humors" and "sweetening the bitter *sucs* [vital fluids]." The most important step was to make sure that, on leaving the bath, you applied Eau Cosmétique to the body with a fine sponge, thus "preventing the body from losing too many of its forces following the action of the bath."[5]

The art of perfumery was entering a new age. It was time, they said "to join a little theory to a lot of practice."[6] The youngest member of the household, Édouard, took the dictum to heart. The next year, he left home to try to become a chemist.

PARIS, RUE SAINT-JACQUES ON THE LEFT BANK, 1825

Édouard was nineteen when he left, crossing over to the Left Bank of the Seine to settle in Paris's Latin Quarter. He joined a group of young men surrounding Louis Jacques Thénard, the former prize student of Fourcroy and Vauquelin, who was rapidly coming to dominate the field. By this time, Thénard held the three most important chairs of chemistry in France, at the Collège de France, the Sorbonne, and the École Polytechnique. He benefited from a strange French system called *cumul,* which allowed academics to hold several positions simultaneously, which meant that one person might control access to an entire discipline. Thénard was not the only one to do this. Biot, who had held the chair in physics at the Collège de France when he set off to investigate the meteorite at l'Aigle, had picked up the chair at the Sorbonne in the intervening years. Lately, though, he had not been seen much at either location, having assigned his classes to substitutes. Thénard, meanwhile, had never been more popular. His lectures were theatrical affairs, drawing not only students but an interested public, who were thrilled by the explosions and colorful displays. The crowds were mostly men and some women, and ranged from pharmacy students learning their first chemistry to independent citizens curious about the latest discoveries.[7]

Charles X, restored to the throne of France after Napoleon's defeat, had just granted Thénard the title of baron.[8] It was a throwback to the prerevolutionary days of aristocratic privilege, but it also captured something of the relationship he had with the young group of rising chemists around him. He controlled access to resources and doled out favors and positions with the power of a feudal lord. He held court at his *atelier de chimie*, a large workshop at the Collège de France across the hallway from the amphitheater where he lectured.[9] The school had just expanded the chemistry rooms to take up nearly the full left wing of the building. The workshop was for designing and rehearsing the experiments he would perform in his demonstrations. But Thénard's work there with his assistants also served as a new form of apprenticeship, where he helped them hone their craft.

Thénard presided over a large network of assistants, preparers, and acolytes. There were the *préparateurs,* who helped prepare the materials he would use in his presentations. There were the *répétiteurs,* who went over the lessons with the students after the lectures. Most senior among them was Louis Le Canu, a sturdily built twenty-five-year-old who had been the *chef des travaux chimiques* since 1822. He was the only one allowed to accompany Thénard in the amphitheater during demonstrations, and he practically lived in the chemistry workshop, where he recalled Thénard giving them dancing lessons with chemistry equipment as partners, and laboratory accidents requiring quick trips to the pharmacy next door.[10] Since the field was so new, few of them arrived with degrees in chemistry. Rather, most of them, like Édouard, came from a family trade that

FIGURE 18. Louis Jacques Thénard, surrounded by his students at a chemistry lecture at the École Polytechnique.

was chemistry-adjacent. Le Canu was the son of a pharmacist, as were many of the young men who sought out Thénard. There were also a number of others whose professions were loosely connected with the craft of apothecary, such as doctors, industrialists, and a wealthy mineral water producer.[11] The son of a perfume manufacturer was hardly out of place.

Thénard's students came from all over Europe. One of the most zealous was a young German, Justus Liebig, who had been the son of a hardware merchant. When he was a boy, his father had tried to apprentice him to a druggist, but when he could not afford the indenture fee, Liebig was sent home. The boy continued studying chemistry on his own, using old, out-of-date pre-Lavoisian treatises like Pierre-Joseph Macquer's *Chemical*

Dictionary and Georg Stahl's *Phlogistic Chemistry*. This was enough to earn a grant to study in Paris, and once he arrived in 1823, he approached Thénard as, he recalled later, an awkward young student with "no other introduction to you except his love of study."[12]

There was Jean-Baptiste Dumas, who arrived from Switzerland that same year. He had been apprenticed as a boy to an apothecary in the south of France, but he left the position in frustration, walking on foot all the way to Geneva.[13] There, he took up the study of science, attending classes at the university and reading on his own. (Biot's textbook, *Traité de physique*, was a particular favorite.) When he happened to cross paths with the world-famous naturalist Alexander von Humboldt, Humboldt urged him to study chemistry in Paris. He arrived there at the beginning of 1823, and by December of that year, after a little luck and a lot of reliance on the kindness of strangers, he snagged a coveted spot as one of Thénard's *répétiteurs*.[14]

This was the circle that Édouard entered. Although he did not have one of the official assistant positions, he was part of a larger network of supplemental instruction. He ran a lab of practical chemistry described as "authorized by the University and honored by the goodwill of M. le baron Thénard and many other illustrious professors." It was located at rue Saint-Jacques, n. 41, right where the University of Paris and Collège de France sat facing each other. The courses were set up so that, immediately following one of Thénard's lectures, the students would head to the lab and have a chance to try out for themselves the procedures they had just heard described. The papers reported that the classes became very popular over the course of three years, with both French and foreign students who had come to Paris to study chemistry.[15] Édouard had to hire a new assistant help him: Anton de Kramer, another young man in his early twenties. Born in Milan of German ancestry, Kramer had studied chemistry in Germany and Geneva before ending up in Paris, studying under Thénard at the Collège de France.

Édouard taught his students, who were mostly going into either medicine or industry, practical techniques like distilling and purifying, as well as a wide range of ways to identify various compounds. He gained some renown for his techniques, and in 1828, he and Kramer published a work called *Synoptic Tables, or a Summary of Chemical Characteristics of Salifiable Bases,* which listed tests that could be performed to identify bases that could combine with an acid to produce salts.[16] While the term *salifiable bases* may seem abstruse or recondite, it covered the group of compounds—the alkaloids, including opium, nicotine, caffeine, and other compounds—that had such a clear effect on human physiology. Édouard's methods of preparing volatile alkalis became widely adopted, and were praised in the 1830 textbook of Anselme Payen, a chemist who had studied with Vauquelin.[17] By this time, Édouard had grown more ambitious, publishing a new system of chemical nomenclature that simplified the naming of acids, alkalis, and the salts that they formed.[18]

Édouard also joined a number of Thénard's other students and assistants in contributing articles to a new journal, the *Bulletin universel des sciences et de l'industrie.* Their job was to scour the academic journals and write up short summaries of interesting pieces. Édouard showed a particular interest in the relationship between organic and inorganic chemistry. Chemists still struggled to explain the vegetable kingdom with the clarity and precision they had achieved in the mineral realm. But he was drawn to the parallels between them, reviewing how bodies *"of organic origin,"* like indigo, compared with "simple bodies."[19] Édouard also conducted his own original research. He submitted one of the papers he wrote together with Kramer, "On the Influence of Organic Substances on the Chemical Characteristics of Mineral Salts," to the Academy of Sciences. A two-person committee reviewed it for the rest of the members, praising its work. The main conclusion, they pointed out, was that one should be very cautious when using organic substances, as their "singular" and "extraordinary" effects were hard to generalize.[20] Chemistry was still far from conquering the living world of plants and animals.

PARIS, RUE BOURG-L'ABBÉ, 1828

By 1828, the rue Bourg l'Abbé was looking more modern than ever, with a covered passage of glass and wrought iron connecting the street to the rue

Saint-Denis. That same year, the first bus line in Paris began, with eighteen-passenger horse-drawn carriages that passed directly in front of the shop.[21]

But there were clouds on the horizon for Laugier Père et Fils, foremost among them a new rivalry with England, where industrialization was running well ahead of France. Laugier had always relied on a strong foreign export business, and the house's reputation went well beyond France's borders. As far away as Stuttgart, the novelist Wilhelm Hauff chose their establishment as the example of a well-stocked cosmetics boutique in his 1826 book *Freie Stunden am Fenster*.[22] For over a century, France had been unparalleled in its position as Europe's greatest purveyor of perfumed soaps, powders, and pomades. But this long history, so closely intertwined with the absolute monarchy and the perfumed court of Versailles, came with a downside. The scented hair powders and pomades, the thick, chalky pastes full of lead and arsenic, the heavy scents that lingered in the air—all were now completely out of fashion. Consumers were looking for something fresher and more natural. And they were starting to look to England.

A London barber, Andrew Pears, had developed a gentle, transparent soap from glycerin, scented like "an English garden," that was gaining in popularity. Most alarmingly, it seemed popular among the Americans, an enormous market just beginning to be tapped. Even in the heart of France, the English soap made its inroads. A new perfumer, Pierre-François Pascal Guerlain, opened a store on rue de Rivoli in 1828.[23] Although born in France, he had left for England at nineteen to learn chemistry and soapmaking. After completing his training there, he returned to France, promising his customers the most modern and scientific techniques.

Laugier Père et Fils was not about to be outdone. The job called for a chemist, and Édouard willingly obliged. After some experimentation, he perfected a technique of adding equal parts of tallow soap and alcohol in an alembic, heating it very gently, and collecting the volatilized product. He then recondensed it into liquid and poured it into molds. After three weeks of drying, his bar of transparent soap rivaled that of Andrew Pears.[24] It was every bit as gentle on the skin and had an even finer and more delicate scent. Before long, they were exporting it back to England, and the house began advertising "English spoken" at their establishment.[25] The press celebrated the achievement as a win against England, singling out Laugier in an article outlining how beauty products used to cause damage to the skin, but now, through informed manufacture, offered no danger. "Perfumery," they announced, "has made the most fortunate progress since the revolution. It owes it to chemistry."[26]

As one crisis was being resolved, another one loomed. This one concerned another specialty of the Grasse region: orange-flower water, a fragrance that, like rose water, often doubled as a cooking ingredient. The problem began in 1829, when the medical commission of The Hague put out a notice stating that "Orange blossom water, especially that coming from France, was harmful and dangerous because of the ingredients containing lead."[27] They accused the French of purposely adding a lead compound as a preservative, despite the negative consequences to public health. The notice circulated around the Netherlands, and orders for French products plummeted.

Laugier Père et Fils had run a strong export business in orange-flower water out of their depot in Grasse. They watched the orders dry up, despite being quite sure that they had never at any point added anything to it, let alone lead. Édouard's mother called him back to the shop, where he tested all the orange-flower water there and found it safe. He then approached the Paris *Conseil de Salubrité*, or Health Council, and, with their permission, tested the orange-flower water of a great number of pharmacists, perfumers, druggists, and spicers. Shockingly, many of them did indeed have high levels of lead. The Health Council recognized the extent of the crisis, noting, "not only was the public health in play, but the prosperity of the Midi found itself compromised."[28] Two members of the council, the chemists Pierre-Joseph Pelletier and Antoine Labarraque, joined his efforts, and they tested every vendor they could find. The mystery slowly unfolded: the vendors whose product tested pure always shipped it from Grasse in special glass bottles called *sacoches*. The others shipped it in thin copper vases called *estagnons*. These vases, the investigators concluded, were being lined or soldered with lead. Pelletier wrote recommendations to solve the problem, acknowledging in his report the role of "this young chemist," Édouard Laugier.[29]

That same year, Édouard was asked to contribute to a new encyclopedia of technology and science. The idea was to update Diderot's eighteenth-century *Encyclopédie* with all the new technologies that had appeared since then: steam engines, parachutes, hot-air balloons, and so on. Diderot had subtitled his work *The Universal Dictionary of Arts and Crafts*, and this new version was called, with little subtlety, *The New Universal Dictionary of Arts and Crafts*, or the *Technological Dictionary*. A "society of savants and artisans" edited the volumes, and they brought together France's top scientific minds as contributors.[30]

The revision of the encyclopedia was prompted by anxiety. The French feared falling behind the English in the process of industrialization and had been watching the situation with consternation. British chemists, they all agreed, were horribly backward. Just look at some of their most famous examples—Humphry Davy and Michael Faraday—experimentalists who had barely completed any formal schooling and could scarcely keep up with theoretical discussions. Yet while France had been dealing with the Revolution and Napoleonic wars, England had been advancing by leaps and bounds. In the last five years alone, its engineers had invented electromagnets, cement, and blast furnaces capable of producing high-quality pig iron and steel. Steam engines were everywhere, pumping water from mines, powering textile mills, and propelling the locomotives that now connected most of England. As Britain's productivity soared, the *Technological Dictionary*'s editors noted from across the channel that "nowhere has the study of the Sciences been pushed as far as in France; but concentrated in too small a number of heads, their utility has remained restricted and the nation has thereby gained more glory than profit."[31]

They began publishing in 1822, rolling the volumes out over several years like the original *Encyclopédie*. By 1829, they had published volume 15—O through P—which contained Édouard's contribution, a twenty-two-page article titled "Parfumeur."[32] He framed the state of the perfumer's art as poised between ancient, traditional techniques and a rapidly transforming world of chemical industry.

Édouard divided the world of perfume into two halves: Grasse and Paris. Grasse retained an enduring position from perfume's earliest days. It was, he said, "a natal soil for aromatic plants" whose hills spilled over so abundantly with flowers that they often grew without cultivation. He summarized the various techniques used—infusion, enfleurage, distillation—stressing how little they had changed over the years. He offered a special glimpse into how Laugier's Grasse workshop made rose essence, emphasizing the extraordinary labor required to reduce hundreds of pounds of roses into a few ounces of essence. Perfumery was, he admitted, almost uniquely anti-industrial. "It seems not to want to profit from the discoveries which each day, by their progress, the physical sciences are contributing to industry."[33]

In Édouard's description, Paris was more dynamic. Perfumers had been building a catalog of distinct scents, and he was there to spill the secrets. He gave detailed recipes for both classic perfumes and new innovations, such as Guerlain's Eau de Miel d'Angleterre (a "honey water" that contained no honey, but instead replicated its smooth, dusky sweetness with a combination of benzoin resin, vanilla, musk, rose, jasmine, and orange blossom). He described how to make "eaux d'odeurs," which were generally 24 percent fragrance, and "extraits d'odeurs," which were at least 28 percent. He added some of his own family's recipes, including their signature Eau Laugier, as well as an Eau Anti-pestilentielle, which he recommended to get rid of any disagreeable odors left after cleaning with bleach.[34] The article also ran through the merchandise of a well-stocked perfume shop: face creams, rouge, blanc, sachets, diffusers, mouthwashes, tooth powders, and even techniques for curing toothaches that included sulfuric ether, a powerful anesthetic and narcotic.

The *Technological Dictionary* sold well and was a major contribution to French industrial knowledge. But Édouard soon found an even broader audience when his work appeared in the celebrated "Roret manuals" found in thousands of French homes. They were the brainchild of Nicolas

Roret, a Parisian publisher who had started out producing encyclopedias on natural history, but found that what really sold well were popular volumes filled with breezy advice and hands-on instruction. He went on to publish over 300 of these, on topics as varied as beekeeping, firefighting, steam engines, the dangers of onanism, and much more.

He had a stable of authors producing them, but the great star among them was a woman who had adopted the nom de plume of Madame Celnart. It was a bold name, synonymous with taste and style, although the woman who bore it was herself more a figure of awkward bookishness. She was born Élisabeth Canard, a late-in-life child of a provincial mathematics professor. He had poured all of his love of science and history into her, stunned by her "precocious intelligence."[35] By thirteen, she had read all forty-four volumes of the Comte de Buffon's monumental *Histoire Naturelle* and soon joined the boys as a rare female to receive a secondary education. It was an idyllic childhood on the banks of the Allier River in central France, but it ended abruptly when she turned eighteen. Her father moved the family to Paris in 1814, inspired by Chaptal to invest all of his money in a sugar-beet factory, only to see it ruined with the fall of Napoleon's regime. Illness affected the family as well, taking Élisabeth's mother and leaving the girl deaf for the remainder of her life.

Paris may have been the site of many young women's dreams, but she was hardly in a position to enjoy the city's charms. Financially ruined, with no connections and hampered by her deafness from taking part in the repartee that drove the social scene, she spent every evening alone in her room, writing. Her first efforts, romance novels, were described as having "more verve than style," but she kept writing, and her scope broadened to include serious social commentary on the impact of machinery on the labor force.[36] She caught the eye of Nicolas Roret, who recruited her to write the manual for "domestic economy"—the sort of self-sufficient household management for which her modest rural background prepared her well. The manual contained tips on cooking, gardening, and preserving, as well as instruction for making wine, beer, absinthe, Malaga, and kvas (a drink from fermented bread). She ultimately wrote twenty volumes for Roret. Her next two manuals were similarly down to earth—one on raising domestic animals and one on curing meats. But they soon took a more rarified turn, covering topics such as How to Be Good Company, Training Domestics, Artificial Flowers, and Parlor Games.

By 1833, her transformation into France's final arbiter on style and grace was complete when she presented the *Treatise on Elegance.* It was a revision of an earlier work, *The Ladies Manual,* which aimed "to rehabilitate the art of *la toilette.*" Baths were back on the agenda. She recommended spending at least an hour and a half in a bath of cool water once a week, wearing a dressing gown. She gave strict directions for the various scrubbings involved, allowing for closing one's eyes if modesty demanded. She gave up-to-date recommendations on hair care, emphasizing that one should avoid washing hair at all costs, but should instead clean it by a thorough combing, first with a wide-toothed ivory comb, then a fine-toothed one, and finally with a square brush that used either the softest horse hair available, or else the roots of rice plants. She continued her tour of skin care, teeth cleaning, and cosmetics, emphasizing at every point the advantages of incorporating fine scents into the routine, from dabbing eau de cologne behind one's ears to perfuming the hair. (She recommended Huile Antique in the winter and pomades in the summer, with heliotrope, rose, and narcissus the favored scents.) She had strong words against what Linnaeus called the "ambrosial odors": the strong animalics of musk and ambergris. "Many flee before the musked and ambered woman as from a plague victim," she warned, and steered her readers to the clean, balsamic scents of iris, violet, and rose.[37]

The primary change she made to her updated *Ladies Manual* was to "borrow," as she put it, from the recent writing of Édouard Laugier. She copied several recipes he had included in his article and gave detailed instructions on the perfumer's art, from almond soap to vanilla pastilles. She limited herself to the most straightforward perfumes, such as Eau de Miel and *esprit de mélisse.* There were so many more classic scents, she conceded, such as the extracts of *fleurs de pêcheurs, de bouquet, de l'eau de mille fleurs, de mousselin.* But these involved such careful preparation of so many different essences that "one would find it easier to procure them premade from Laugier on rue Bourg-l'Abbé."[38]

Édouard became Celnart's true muse the next year, when she undertook a manual devoted entirely to the art of perfume. Roret had published one of these before. In fact, it was one of the very first manuals he put out when he began publishing them in 1825. The radical feminist Marie Armande Jeanne D'Humières Gacon-Dufour had written the first one, encouraging women to take the art of perfumery into their own hands. She warned against the charlatans who pushed their dubious concoctions and the chemists who added harmful ingredients. But things had changed by the time Celnart was writing, and she wanted to recast the relationship between the art of perfumery and the science of chemistry. Chemists' involvement in the nascent perfume industry in the eighteenth century had left foul traces of alchemical potions, toxic to the body and overbearing to the senses, in the mind of the public. The whole enterprise had become so dominated by charlatanism that many thoughtful women, she conceded, had a strong prejudice against perfume, and prided themselves in rejecting it entirely. Yet a new era had dawned, Celnart insisted. Those educated in the new chemistry were now capable of establishing a rational basis for the art of perfume. "But when one sees the perfumer completely reject this mass of recipes that, previously, made up all of his knowledge, and ask of chemistry salutary combinations; when one sees him study primary materials, combine them according to rational principles, abandon this mysterious march that only imposes itself on the ignorant, no pretext of disfavor can remain against this amiable art, which is becoming more honorable and more lucrative."[39]

Who was this perfumer studying primary materials? She made no secret that it was Édouard Laugier, borrowing directly many of the passages he wrote for the *Technological Dictionary*. Her adoptions were so extensive and direct that they would no doubt be labeled as plagiarism today, although by the standards of the time it was a more acceptable form of flattery. Indeed, Celnart herself soon found her work borrowed as well. Campbell Morfit, a chemist working in Philadelphia, published *Perfumery: Its Manufacture and Use* in 1847, noting that it was by and large lifted from the French text by Celnart.[40] The French, he noted, were the masters of the art, and he saw no better way to instruct American perfumers, druggists, and soap manufacturers in the trade. And thus Édouard's technical advice spanned continents as the latest word on scientific perfuming.

PARIS, COUR VIOLET, 1828

By now Édouard had a new job title. He was, as the *Journal de Commerce* called him, the "distinguished professor, M. Edouard Laugier," teaching at a new school, the Lycée Commercial et Industriel.[41] It was the brainchild of a businessman, Antoine Valin-Ponsard, and a philosopher, Joseph Morand, who wanted to establish a school of practical education for a nation "whose entire force is in its industrial class."[42] They arranged courses on economic theory, accounting techniques, and the mechanical arts, being sure to offer class both during the day and at night so that men already employed in industry would be able to attend.

They hired Édouard and Charles de Filière to share the duties for the chemistry section. On Mondays, Wednesdays, and Fridays at three in the afternoon, Édouard would lecture on natural history and applied chemistry, while on Tuesdays, Thursdays, and Saturdays, Charles would give a course on chemical manipulations.[43] The directors had a bit of trouble arranging a location. The first announcements, in 1827, claimed the school would be sharing a space on the rue de Bac.[44] But by the time classes began in 1828, the school had moved across the river to the Right Bank, to the cour Violet in the neighborhood of Faubourg Poissonnière. The announcement described Édouard Laugier as "professor of chemistry, student of M. Thénard."[45]

A school for the industrial class was an excellent idea, perfectly fitted to the times. Justus Liebig, when he had returned home after working in Thénard's lab, had wanted something similar. He had taken a position teaching chemistry at the University of Giessen, and had proposed adding a lab to it that focused on practical, industry-oriented instruction. The University Senate, however, was not interested, responding sharply that it was their task "to train civil servants, not apothecaries, soapmakers, beer-brewers, dyers, and vinegar-distillers."[46] Liebig continued anyway, starting a private institute in 1826 and advertising in pharmaceutical journals to attract students.[47] Because it was outside of the university curriculum, he accepted a wide range of students, including tradespeople who did not have the credentials to attend a university.

Another of Thénard's students, Jean-Baptiste Dumas, also started a school. He had been giving talks at the Atheneum by the Palais Royal, but these were lectures for a popular audience, not a comprehensive curriculum. In his mid-twenties, Dumas looked preternaturally younger, with a round cherubic face and ringlets of soft curls, and his audiences found him charming.[48] One evening, Alphonse Lavallée, wealthy owner of the newspaper *Le Globe*, was in the audience and could not get over the silky polish of the young man. He sought him out and offered to bankroll whatever sort of enterprise he wanted to embark upon. Dumas proposed a school that would teach practical chemistry techniques to the growing commercial and industrial classes. In short, it was precisely like the school where Édouard Laugier was currently teaching, only this one would be flush with money.

Lavallée was not the only rich patron on Dumas's side, as by this time he had married into the wealthy and well-connected family of the Brongniarts. He had first befriended the son, Adolphe, when the two began editing a scientific journal together in 1824. Two years later, he married Adolphe's sister, Herminie, and soon had the occasion to ingratiate himself with his father-in-law, Alexandre Brongniart, who ran the state porcelain works at Sèvres. The new King Charles X, who had assumed the throne in 1824, had turned to Brongniart after a debacle at his latest ball at the Tuileries. When the servants lit the candles in the ballroom, irritating fumes forced the guests to flee, coughing and rubbing their eyes. Brongniart passed the problem on to Dumas, who figured out that the candles had been bleached with chlorine, and when they burned, formed

FIGURE 19. The inner courtyard of the École Centrale, located in the Hôtel de Juigné, a palatial former residence.

hydrogen chloride—a gas that attacked both the eyes and the throat. His father-in-law was pleased that the problem was solved, and Dumas's reputation as a wunderkind was secured.[49]

His new school, the École Centrale des Arts et Manufactures, opened its doors on 3 November 1829. This was more than a year after Édouard's own opening, but its scale far eclipsed the modest enterprise on rue Violet. Lavallée had procured a literal palace for Dumas. The Hôtel de Juigné had been one of the grandest private residences of Paris (and visitors can still tour it today as the location of the Picasso Museum). It was enormous—6,000 square meters of space—and included a stately entrance hall with a sculpted marble staircase and numerous rooms surrounding a private courtyard. And there was money left over to equip three laboratories—two for preparing class demonstrations and a larger one for "industrial chemistry."[50]

Dumas taught the course on chemistry. His brother-in-law, Adolphe, taught natural history.[51] They brought in two more friends to teach mathematics and physics and together the young men wrote a manifesto challenging the "industrial superiority of England."[52] Dumas even hired his own assistant to help with the chemistry course, bringing in Auguste Laurent, a thin, intense twenty-three-year-old who had just graduated from Paris's celebrated School of Mines. He had fallen in love with chemistry while touring the mines of Germany for his thesis. Once back in France, he finished his degree but then abandoned the career of a mining engineer to work for almost nothing as Dumas's assistant. The two kept busy shuttling between the school's three laboratories, as its enrollment surpassed all expectations. In the first year, 140 students attended the courses, many of them older than their professors.[53] And the numbers continued to grow.

The success of the École Centrale far outstripped that of Édouard's Lycée Commercial, draining students away from it. Without enough demand for two industrial schools, the Lycée folded. Charles de Filière went to Rouen and reopened the school there. Édouard moved back in with his parents.

7

Lost Illusions

PARIS, RUE BOURG-L'ABBÉ, 1830

Édouard was the only one home when Monsieur Barni knocked on the door of Laugier Père et Fils at eight o'clock Monday morning, flanked by two clerks of the Justice of the Peace. Although it was the middle of August, the morning was cool, a break from the heat wave that had hung over Paris. The street itself was calm and quiet, although it still bore scars from two weeks earlier, when a dense network of barricades blocked its passage.[1] France had lived through another revolution, this one short but effective. King Charles X had been growing more despotic throughout his reign. His final act, in 1830, was to revoke the charter granting the minimum of basic rights to his citizens, taking particular aim at the freedom of the press. Parisians knew what to do. Over 4,000 barricades blocked roadways across Paris, with two of the largest on rue Saint-Martin and rue Saint-Denis, leaving Bourg-l'Abbé, nestled between them, a haven for the insurrectionists. As an eyewitness in the neighborhood recalled, "Every man was armed. Women were occupied with their children in unpaving the streets, and carrying the great stones into the houses in order to shower down upon and crush the military."[2]

The July Revolution, as it was called, toppled the government in three days, replacing Charles X with a king who promised to honor a constitution and the principle of popular sovereignty. But it precipitated a crisis for Maison Laugier. The challenges of the past couple of years had already left them in a fragile state, with debts mounting to over 33,000 francs in 1828. The revolution sent them over the edge. A substantial portion of their business was with foreign customers, both in Europe and the colonies. Unable to collect on these foreign debts, they declared bankruptcy.

Monsieur Barni and the clerks had arrived that Monday morning to take an inventory of everything the Laugiers owned, as it now belonged to their creditors. Édouard, who was described in the clerk's notes as an employee that lived on the premises, showed them around. They spent three full days together, from eight in the morning until five in the evening. First they inventoried the office and bedrooms on the second floor, noting the material and quality of every piece of furniture, the size of every mirror, the contents of every wardrobe. Next was the wine cellar. They checked the status of the corks on 300 bottles of wine and two casks, then proceeded to the third floor *savonnerie* and counted the bars of soap. The store itself took a day and a half. There were two large rooms, separated by a corridor. In one, they opened each of the twenty-six large cases, noting its contents: "90 bottles of essences for spots," "18 kg almond paste," and so on. In the other room, they noted 500 liters of "various spirits" and 3,000 glass bottles.[3] The corridor itself was lined with perfumed oil and spiritous tinctures. There was a mezzanine with several full armoires and a counter with 2,400 pots of pomade. There was also a back room on the ground floor devoted to brushes, whether for hair, teeth, beards, or clothes.

The very last room they visited was the laboratory in the back, where Édouard now spent most of his time. They listed the inventory of his workspaces: three different kinds of alembics, together with their serpentine coils, four big bains-marie with their copper furnaces, three wine presses, a mustard mill, a cast iron boiler, a winch, a fountain, a pump, three copper basins, a butcher block, two tables—a large and a small. Smaller items did not escape their notice, including an axe, eight oil lamps, a marble mortar, a copper mortar, an iron mortar, three strainers, two big copper tubs, fifty molds for pomade, eight carboys, a balance and set of weights. The list continued as the clerks made their way through the bottles, pots, tins, jugs, and carboys on the shelves. Édouard then informed them of their factory in la Chapelle Saint Denis, and he gave them a list of the equipment and merchandise held there.

The judge in charge of the case ultimately had sympathy for the perfumers. "In this embarrassing circumstance," he noted, it was important to avoid closing the store. Doing so could "destroy entirely a long-standing House known favorably for its manufacture" and "augment the common Misfortune."[4] The judge let the bankruptcy proceed in secret, making arrangements between the Laugiers and their creditors. The business kept operating as usual. It was "indispensable," the judge noted, that "the public notice nothing about the state of bankruptcy in which the house has fallen."

But the story of one of Paris's top perfumers going bankrupt was too good to go unnoticed. It caught the eye of a young Honoré de Balzac, then working as a kind of gossip columnist of the fashion scene. He had just started a new column in the newspaper *La Mode,* doling out advice like "Carelessness in *la toilette* is a form of moral suicide."[5] The plan was to start with suggestions for clothing and then move on to perfumes, bathing, and hairstyling.[6] But after the legal notice of Laugier's bankruptcy appeared in the papers, he abandoned that project to start a novel about a successful perfumer who goes bankrupt.[7] He began the book, ultimately called *The Rise and Fall of Cesar Birotteau,* in 1830, although he put off publishing it for six years, worried his audience might be bored by the "vicissitudes of bourgeois life" it presented.[8]

Balzac presented Birotteau as a pioneer of nineteenth-century advertising, with many of the ads echoing those of Laugier Père et Fils. For example, the fictional perfumer's prospectus for his signature product, the Pâte des Sultanes, read, "It has been approved by the Institute on the report of our illustrious chemist Vauquelin." The widely distributed prospectus of Eau Laugier read, "the precious qualities of the eau adopted throughout all of Europe have been recognized by MM. Deyeux and Vauquelin, following their report approved by the Faculty of Medicine of Paris."[9] Birotteau's best-selling item was an oil pressed from hazelnuts that he claimed could regrow hair and preserve its color, much like the Laugier's recently introduced hair oil, Huile à la Neige.[10]

Balzac ultimately gave the story a happy ending, in which Birotteau paid back all of his debts and rebuilt his perfume empire. And the House of Laugier was able to recover as well, with Édouard staying on to run the laboratory. By 1831, the store was once again running profitably. And it was a good thing, too. France had never been so preoccupied with the demands of good hygiene. A cholera epidemic was making its way west from Russia. It seemed inevitable that France would soon succumb, and soaps and disinfectants flew off the shelves in an effort to ward off the foul smells that carried it.

PARIS, ÉCOLE CENTRALE, 1832

Cholera was a horrific disease. Its sufferers took on a cadaverous aspect, with sunken eyes and shriveled skin that turned blue or purple. They writhed in uncontrollable spasms, punctuated with bouts of vomiting and a peculiar, fishy-smelling diarrhea. It spared no one, killing the young along with the old, the strong along with the weak. It had arrived in Moscow in the autumn of 1830. Within weeks, half the city was dead. From there it began a relentless march westward. Each month brought the announcement of a new city that had fallen prey: Warsaw, then Berlin, then Hamburg.

By the time cholera reached England, in December 1831, France had already begun preparations. Quarantine regulations went up immediately, governing incoming ships and dictating that all letters from England "shall be pierced and passed through vinegar, or an aromatic fumigation."[11] Medical opinion was still up in the air on whether cholera was spread by contagion or miasmas, but both sides agreed on one thing: one should avoid bad smells. The nose remained the best detector of the traces of effluvia, discharge, exhalations, rot, or decay that signaled disease.

The prefecture of police in Paris had, months earlier, written a pamphlet entitled *Popular Instructions* for dealing with cholera, and they now distributed over 40,000 copies throughout the city. The pamphlet gave tips on avoiding unhealthy odors: opening windows, emptying chamber pots right away, not drying laundry indoors, not leaving rotten vegetables about.[12] But the bulk of it was devoted to recipes for remedies to either prevent or treat cholera. Camphor, with its bracingly overpowering smell, emerged as the most important ingredient. The pamphlet recommended camphor-infused steam baths for the sick. It gave recipes for various camphor liqueurs that the afflicted took by the spoonful every fifteen minutes. It instructed the reader to fill a jar with eau de vie, vinegar, camphor, pepper, mustard powder, and garlic, then leave it in the sun for three days to produce a powerfully odorous liniment, which they should then rub vigorously all over the sick person's body.

Camphor had long been known for its overwhelming smell, its sharpness and astringency a sharp contrast to the soft fetidity of decay. "Strong, penetrating, not disagreeable to some people but unbearable to others" was one description.[13] It had lately gained popularity as an antispasmodic, which may explain why it became the favored treatment for cholera.[14] The supply of camphor could scarcely keep up with the new demand. It was made by distilling the bark of specific varieties of laurel trees that could only be found in Japan, Borneo, and Java.[15] Botanists had been trying to get it to grow in France for decades, but had only been able to get a single tree to flower at the Museum of Natural History and had never gotten any camphor from it. All of the commercial product was imported, and it was not enough. In this moment of national crisis for France, Dumas and his assistant Laurent began to explore the possibility of making artificial camphor from more readily available sources.

By the time cholera arrived in Paris in March of 1832, all instruction was suspended at the École Centrale.[16] The classrooms were empty. Dumas and Laurent had the teaching laboratory entirely to themselves, surrounded by the objects of their new research: camphor, peppermint, anise, turpentine, and more. The idea of an "artificial camphor" had been around for a while. The French chemist Joseph Proust had first used the term in 1789. He was living in Spain at the time, in a region hosting what he called an "extreme abundance of aromatic plants."[17] While working with

FIGURE 20. Interior view through a doorway showing the treatment of victims of the 1832 cholera epidemic in Paris.

some of the local lavender he noticed that its essential oil, when left exposed to the air, deposited a substance that had all the characteristics of camphor: solid but flexible, white yet translucent, with camphor's strong, distinctive aroma. Other aromatic plants such as peppermint, sage, thyme, rosemary, and marjoram also produced camphorlike substances.

But were these artificial camphors really the same thing as the natural one? The best chemists of France had tackled the question. Thénard had even developed his own technique for making artificial camphor from turpentine.[18] Also known as the essential oil of terebinth, turpentine was a distillation of the resinous sap of the terebinth tree, which was widespread in the Mediterranean, including the south of France. It was easier to procure than camphor, and in the cholera epidemic it often served as a substitute. But Thénard was able to say only that it looked and smelled like original camphor. Since he could not accurately establish the chemical composition of either one, he could not say for sure.

This was the task that Dumas and Laurent waded into: to determine the relationship between natural camphor and all of its imitators. But unlike Thénard they had a secret weapon, a delicate glass apparatus called a kaliapparat. Justus Liebig had invented it after running into the same problem that had confounded everyone else when analyzing organic substances.[19] When burned, the carbon converted into carbon dioxide gas, which was difficult to collect and accurately measure. To get around this, Liebig had taken a glass tube and introduced a series of kinks and hollow spheres. He filled the center sphere with a caustic potash solution. When he tried to analyze a substance by burning it, the potash solution would react with the substance's carbon dioxide gas (which was hard to measure) and trap it as a solid, potassium carbonate (which he could easily weigh). By subtracting the mass of the kaliapparat before combustion from its mass after combustion, Liebig could calculate the mass of carbon in the original substance more precisely, and so he was able to produce estimates of the chemical formulas of morphine, quinine, and a number of other substances.[20] Dumas was the first to review this work in France, and he marveled at the potential of this new device.[21] He soon had a chance to use one himself. One of Liebig's students, Charles Oppermann, brought a kaliapparat to Paris in December 1831, and Dumas invited him to his lab.[22] After analyzing some camphor with the kaliapparat, Dumas wrote a brief but enthusiastic note to the *Annales* announcing that "the moment is not far off where the majority of organic substances will find themselves methodically classified on the same basis as in mineral chemistry."[23] The messy world of living things would soon, he believed, be tamed.

As cholera spread in the spring of 1832, Dumas worked double time to solve the question of artificial camphor. "I was in a big rush," he admitted, "to compare the camphor of lavender that existed in the collection of the Collège de France with ordinary camphor."[24] He tallied up the number of carbons, hydrogens, and oxygens in both, and had Laurent verify all the results. The artificial version, they found, gave "exactly the same results as ordinary camphor."[25] It was exciting stuff—he was the first chemist in France to use the kaliapparat to successfully analyze organic materials, and his work indicated that camphor, so high in demand, could be made artificially from other, cheaper ingredients.

All around him, cholera continued its relentless course. The disease struck both the low and the mighty, including some of the most prominent names of science. André Laugier, who had succeeded Fourcroy at the Museum of Natural History and Vauquelin as the director of the School of Pharmacy, was one of the early victims, dying on April 19. Georges Cuvier fell ill soon after. He was perhaps the most important figure in French science at the time, having served for years as the permanent secretary of the Academy of Sciences for the physical sciences, the highest position of the most important scientific body of France. His funeral was a stately affair, involving a procession from the Jardin des Plantes to the church, and orations by representatives from all four academies of the Institute.[26] Unfortunately, it turned into an occasion for the disease to spread. Academy member Georges Serullas caught it while attending, and, along with other Academy members such as Sadi Carnot, Henri Cassini, and Jean-François Champollion, soon passed away himself.

Serullas's death left an opening in the chemistry section at the Academy of Sciences. Dumas wanted it, and together with his mentor Thénard, he began to wage a subtle campaign. The way the academy worked was that only members were allowed to present publications known as "memoirs," but they could also give "reports" on the work of nonmembers. Starting in June, Dumas began submitting memoir after memoir, and Thénard began to give weekly "very favorable" reports on his work.[27] Thénard was also at the head of the committee that compiled the list of candidates to replace Serullas's seat. Dumas's name was first among the five candidates listed, and when the vote took place, on 6 August 1832, he won with an overwhelming thirty-six of the forty-four votes cast. He immediately became one of the most active and involved members, attending every session and submitting his own memoirs with regularity.[28]

It had been a dizzying few months. His work on camphor had launched his career. The ability to precisely analyze the chemical formula of natural camphor and prove it was identical to artificial camphor seemed to be an uncontestable triumph. Until, that is, one of his childhood heroes returned from obscurity to throw all of his claims into suspicion.

NOINTEL, FIFTY MILES NORTH OF PARIS, 1832

On a clear spring day in May 1832, as the rest of the scientific community of Paris made preparations for Cuvier's funeral, Jean-Baptiste Biot found himself fifty miles outside of the city crouched on the ground, trying to light a flower on fire. It was no easy task. He had spent fully two years perfecting his technique. But he was a patient man, with flowers if not always with human beings, and his efforts were beginning to pay off.

The flower in question was a fraxinella, a beautiful thing with a surprising lemony smell that left a lingering warmth of cinnamon. But what made it famous were the flames that engulfed it when ignited, then subsided to leave the flower unharmed. It was, the story goes, Linnaeus's daughter who first noticed the effect when she approached the flower with a candle to get a better look. Linnaeus himself had tried to recreate the effect several times, but always failed, as had many people after him. The flower's strange ability took on a near mythic status—dismissed as "delusion" by some, equated with the elusive "*esprit recteur*" by others.[29] Biot himself suspected "the action of life" was behind it, which was what brought him back, again and again, to kneel before the flower.[30] On his first try, on 26 April 1830, he was rewarded with only a little sputtering of flame, reminding him, he said, of what happens when you squeeze the peel of an orange near a candle. But he found that it became easier as the plant matured. By 1832, he had mastered the effect.

It was a delicate plant that craved warmth, more common in Provence than north of Paris, but Biot insisted on having it in his garden and, with persistence, had managed to grow two different varieties, one with white flowers and one with red, which he found gave off a bigger flame. He spent a lot of time in his gardens these days. They had once been part of the grand estate of a marquis in the *ancien regime,* in whose kitchens, legend had it, Béchamel sauce had been invented. The estate was in ruins by the time Biot bought it, left demolished by the Revolution. The chateau was little more than crumbling rubble, so Biot and his family lived in one of the former outbuildings, off to the side.[31] The gardens had once been spectacular, with charms that would rival Versailles: flowing fountains, a labyrinth, fruit groves. Although they, too, had fallen into ruin after the Revolution, they were, for Biot, both his refuge and his laboratory.

It had been nearly thirty years since Biot had first set off for l'Aigle to investigate meteors for Fourcroy and Vauquelin. And what a strange, tumultuous journey it had been. In 1804, he rose over 4,000 meters in a hot air balloon, risking the thin air and cold to become the first scientist, with Joseph-Louis Gay-Lussac, to conduct high-altitude experiments.[32] In 1806, he traveled to Ibiza and the mountains of Spain on an expedition to measure the meridian. The Academy of Sciences had sent him and a partner, needing an accurate measurement for the new metric system, which defined the meter as one ten-millionth of a meridian quadrant.[33] Halfway through the scientific expedition, however, Napoleon started a war with Spain, and suspicious locals mistook their activities, which involved signaling to each other across mountain peaks with lit fires, for spying. Biot barely made it back to Paris safely. His partner, François Arago, was not so lucky. Arago, who was assumed dead, surfaced more than a year later to regale people with tales of being chased through town by angry Spaniards, breaking himself out of prison, sneaking through North Africa

FIGURE 21. Joseph-Louis Gay-Lussac (*left*) and Jean-Baptiste Biot (*right*) in their 1804 balloon ascent. They rose 4,000 meters to conduct the first scientific experiments at these altitudes.

dressed as an Arab, being held hostage by the dey of Algiers, and being captured by pirates before making it back to France.

Back in Paris, Biot continued to be a golden boy of the new age of science, moving with his wife and son into an apartment above the Collège de France. He was a particular favorite in the group of men that clustered around Pierre-Simon de Laplace and met every weekend at his large estate outside of Paris. Laplace had never forgotten the visits in his youth to Lavoisier's home at the Arsenal, learning more from overhearing his conversations with Fourcroy than from any classroom. He hoped to recreate the experience for a new generation.[34] His goal was to reform physics the way Lavoisier had reformed chemistry, giving poorly understood concepts like light, heat, electricity, and magnetism a rigorous experimental foundation. In 1809, he and his group had a brand new topic to discuss: "polarization," a term just coined by one of their members, Étienne-Louis Malus, after noticing something odd about light that reflected off a glass windowpane.

The odd effect echoed an aspect of light that had troubled Isaac Newton since the seventeenth century. When light passed through a special kind of crystal known as Icelandic spar, it broke into two beams of complementary colors, which were called an ordinary and an extra-ordinary ray. If one took a second crystal, placed it on top of the first, and rotated it, the beams would vary in intensity and perhaps even disappear. This led Newton to ask, "Have not the Rays of Light several sides, endued with several original Properties?"[35] Perhaps, he speculated, light had a long side and a short side, like a rectangle. When a beam of light passed through the Icelandic spar, all of its rays got their sides lined up in the same direction. As the beam went through the second crystal, if the long-sided light rays tried to pass through the short-sided slots, the beam disappeared. Malus had noticed the same thing happening with the light reflected off a glass window. He could make the beam diminish and disappear by passing it through a crystal, as if all the light's "sides" were lined up in one direction and passing through the slots the wrong way. He called a beam of light with its sides aligned "polarized," and began a number of experiments to try to work out the mathematical laws of what was going on.

When Malus died in 1812, weakened from a bout of the plague he had picked up when accompanying Napoleon on his expedition to Egypt, Biot took up the topic. Within a couple of years, he had a discovered yet another strange quirk: when he passed polarized light through certain crystals, such as mica or quartz, the direction of the "sides," which he called the plane of polarization, rotated a few degrees, as if somebody had given a turn to a steering wheel. He named this new thing rotary polarization, usually translated to circular polarization in English. He guessed that there was something about the crystalline structure that twisted the light as it went through, and he started fiddling with different thicknesses of crystal and putting them at different angles to see how that changed it. He was using a device called a "polarimeter" that consisted of a metal tube about three centimeters long. Before the light went into the tube, he would polarize it (in the early models, this was sometimes accomplished by reflecting light off a glass plate, but it was ultimately easier to use a piece of Icelandic spar). He then placed the quartz or mica crystal in the tube and observed what came out the other side. As his samples got thinner, they got harder to position, and he decided to try filling the tube with a clear liquid, as it might make it easier to keep them suspended inside. One of the liquids he used was turpentine, and this led to the observation that would haunt him for the rest of his exceptionally long life.

Biot noticed that the turpentine by itself, before he even placed a crystal in it, was somehow rotating the plane of polarization of light. The effect was slight, but Biot tried to amplify it by employing a longer tube, of sixteen centimeters, so the light would have to travel farther through the turpentine. The effect was now undeniable, although it remained shocking. Unlike crystals, which were organized in a lattice arrangement, turpentine was just a normal, homogeneous fluid with no structure to it. He began to test other essential oils to see if they had the same property. They did—including the essential oil of bay tree, the essential oil of lemon, and the dissolution of natural camphor in alcohol. Whatever force was turning the light, Biot concluded, must have been coming from the molecules themselves.[36] He gave this the name *molecular rotary power*, usually called optical activity in English.

He figured that if rotary power were a feature of the molecule itself, then it should be present in the gaseous state as well as the liquid one. He wanted to run an experiment to test this hypothesis but calculated he would need the tube to be at least thirty meters long to see the effect, and few rooms would be large enough to house such an apparatus. Laplace stepped in to pull some strings. He was a longstanding member of the House of Peers, which was housed within the Luxembourg Palace.[37] In 1817, he arranged for Biot to perform his experiment in the Orangerie of the palace, used for housing orange trees in the winter. Biot set up shop in the summer, building his enormous, light-tight tube raised high off the ground by extensive wooden scaffolding. He placed a large boiler at one end to heat the turpentine into a gaseous state. When everything was ready to go, he brought in two assistants to help with the observations.

The turpentine vapor was just growing dense enough for Biot to begin seeing results when everything went horribly wrong. Both the gas and liquid turpentine ignited, exploding the boiler and sending its lid flying across the room. Biot and his assistants watched in horror as a column of fire stretched from the boiler to the ceiling. The wooden planks above their heads caught fire, well beyond their reach. There was nothing to do but alert the fire department, who luckily were nearby and able to contain the flames to the interior of the building. No one was hurt, and the authorities chalked up the damage to "the unforeseen consequences of scientific research" and did not charge him.[38] But it was still a disaster. His equipment lay in ruins. Laplace was angry. And the Luxembourg Palace was definitely off-limits for any future experiments.

Nearly burning down the House of Peers was not the worst event in Biot's life at that moment. He had become, rather unwillingly, the public face of the losing side of the most high-profile scientific debate of the time. The topic was whether light was a particle or a wave. Most physicists from

Newton through Laplace had thought of it as a particle, and Biot had borrowed that language in his early papers, referring at times to "molecules" of light. This was uncontroversial at first, but it became a problem when a young upstart, Augustin Fresnel, proposed that light was actually a wave, oscillating in an up-and-down motion. Polarization, in this case, meant that all of the oscillations were aligned along a single plane. It was a simpler, more elegant explanation than Newton's elongated "sides," but it threatened the Newtonian program Laplace and his students were advancing.[39] They shoved Biot onto the front lines, and he spent the next ten years fighting with Fresnel on the subject. He was a rather unwilling champion of the cause. He admitted that he did not particularly care about whether light was a particle or a wave, and it did not make any difference to the work he was doing on optical activity. But it was hard to break out of the narrative in which he was the chief spokesperson for what was becoming the losing side.

The true extent of his fall from grace became clear in 1822, when the position of permanent secretary of the mathematical sciences opened at the Academy of Sciences. Biot wanted the position badly, and hoped that Laplace's support would help garner him votes. But Laplace refused to give it, and Biot took the subsequent loss as a final humiliation. He effectively dropped out of the scientific profession. Although born and raised in Paris, he left the city entirely, purchasing a ruined castle in Nointel, along the Oise River. This was where he had been sent to teach high school, in the years before meeting Laplace, and where the family of his wife, Gabrielle Brisson, still lived. He described his new home as "a small rural property that I cultivated myself." He embraced rural life, even becoming the mayor in his little village of about 200 people. He stopped publishing. He avoided the weekly meetings of the Academy of Sciences.[40] Although he kept his academic positions, he arranged for replacements to fill in for him as often as possible.

Back in Paris, Biot's colleagues wondered if he had had a mental breakdown and assumed he had given up science entirely. But he was working as hard as ever, and had returned to what he felt was the most important question he had encountered in his work: Why did the essential oils of certain plants rotate the plane of polarization of light? He was convinced that this optical activity had revealed a level of molecular organization of these compounds—an organization unique to living things. His instruments, in short, allowed him to see an aspect of life itself. He packed them carefully and took them to the countryside, where he would be able to observe life *in situ*, as it grew, developed, and took on the marvelous organization that he was intent on revealing.

Through the length of the growing season, from the first peek of buds in the spring to the last of the fruit rotting on the vine in the autumn, he tested the plants of his orchards, fields, and gardens. He squeezed and pressed any bit of fluid he could obtain from them—the sap from trees, the juices from fruits, the fragrant oils of herbs and flowers, the gums and mucilage of stalks. These all fell into the category of *sucs vegetaux*, or the organic fluids that Chaptal had identified as the vital organizing force of plants. In search of this vital force, Biot filled the cylinder of his polarimeter with the various fluids, sent a beam of polarized light through, and recorded how much it rotated. He watched in amazement as the polarimetric readings grew stronger over the course of the spring, as the plants grew and matured, and then dropped once again as they withered and decayed.[41] It was as if he were able to chart the very progress of life.

The revolution of 1830 broke some of the calm of Biot's pastoral idyll. Local villagers ransacked his estate, incensed by what they thought was a portrait on the walls of the hated King Charles X (it was, Biot pointed out, actually of Laplace).[42] The shift in political winds blew him out of office, and he was abruptly dismissed from his position as mayor. Biot began spending winters back in Paris, returning to his apartments at the Collège de France to find that Thénard's cadre of chemistry assistants had sprouted like mushrooms beneath him. This turned into a boon for Biot, who was not much of a chemist himself and relied on them to provide samples of new things to test. Thénard's *préparateur* at the Collège de France was a young man named Jean-François Persoz, whom Biot pestered frequently for various preparations of sugars and starches. Persoz prepared a new substance for him that Biot named "dextrin" after its unusual ability to rotate the plane of polarization to the right (making it, in Biot's language, dextrorotatory). Dumas and Édouard Laugier were also favorite sources, providing exceptionally pure preparations of bergamot, lavender, bigarade, marjoram, anise, fennel, rue, caraway, mint, rosemary, sassafras, savin juniper, and more. Biot dissolved resins in alcohol and heated the waxy essences called "concretes" past their melting point. He poured essential oils directly into the tube of his polarimeter when he had enough. In some cases, such as the oils of neroli and citron, the materials were too precious to fill the entire tube, and he had to content himself with shining light through a glass bottle of the substance.[43] The more materials he tried, the more the picture came into focus: every living thing, he realized, had the power to rotate light—with a sole exception, the essence of bitter almonds.

After ten years of silence, Biot showed up at the Academy of Sciences in October 1832 carrying hundreds of pages of his observations. Reading his work took nearly the entire session, and he returned on 5 November to finish. Optical activity, he argued, was present in two kinds of substances: quartz crystals, where it was due to the organized, aggregate structure, and organic fluids, where it was due to the molecules themselves. The implication, for Biot, was that the organic molecules must have some form of internal organization. Chemists had no way to access this organization, as their analytical techniques destroyed it by separating the constituent elements. Polarized light, Biot argued, was the only way to explore this aspect of living things.

He called his work "Memoir on circular polarization and its applications to organic chemistry."[44] The phrase "organic chemistry" was a new one, a replacement of the old categories of vegetable and animal chemistry. In fact, there was only one person in the audience who was fully a member of this field, Jean-Baptiste Dumas. Yet he was not happy with what he was hearing. Dumas had just been elected to the Academy on the basis of his work showing that artificial and natural camphor were exactly the same, but as Biot continued, he made clear that this claim was completely wrong. He had procured samples of natural and artificial camphor, as well as turpentine and lemon essential oil, directly from Dumas, making sure to specify that these were precisely the same ones that Dumas had used in his recently published piece. Natural camphor rotated the plane of polarization 17 degrees clockwise. Meanwhile, artificial camphor made from turpentine rotated the plane of polarization 24 degrees counterclockwise, which was, Biot pointed out, the same as turpentine itself. He also tested some of the artificial camphor made from the essential oil of lavender and found that it was utterly inert, unable to affect the light at all. The dissimilarity between the camphor that came from a tree and the one that came from lavender oil was, he said, "indubitable."[45]

It was quite a blow from the man whose books Dumas had studied as a boy. Dumas sprang to defend himself, dismissing Biot's distinctions as "inexplicable isometries, with no apparent cause." The concept of isomers was a recent one, gaining traction through the work of German chemists Justus Liebig and Friedrich Wöhler. The two had been studying substances with identical chemical compositions, but Liebig found his to be explosive while Wöhler's were not. They argued a bit before realizing that even though their two molecules had the exact same atoms, they had different chemical formulas that arranged these atoms in different orders, and they began to use the term "isomer" to refer to this. But Biot was going much further. He claimed the substances had the same chemical formula and the same chemical properties, and the only way to distinguish them was by looking through an instrument that he alone possessed. Dumas soon became Biot's chief critic, intent on his efforts to, as he put it, "reject with ample energy the introduction of these vague isometries into science."[46]

There was a difference, Dumas said, between "organic substances," which were what chemists studied, and "organized substances," which took into account the complex arrangement of living things. Biot had stepped over the line into the study of organization itself, passing from the study of "the constitution of chemical compounds" to the "much more hidden constitution of truly organized products, with which Chemistry refuses to seriously occupy itself at the present moment."[47] Dumas was clear in his position: all that mattered was the chemical composition, and an artificial substance with the same composition as the natural one was no different. The chemistry community followed his lead, and most people wound up ignoring the cranky old man with the strange instruments. Only one chemist in Paris, Auguste Laurent, would heed Biot's call, and it would very nearly cost him his career.

PARIS, ÉCOLE CENTRALE, 1833

Laurent was as dreamily otherworldly as Dumas was ambitious. His parents, wine merchants in eastern France, had despaired at how little of the family's business sense he inherited. Reportedly, as he was growing up, his father would leave him to watch the counter only to find that Laurent had wandered off to go exploring.[48] He spent hours in the meadows and marshes surrounding the tiny village of La Folie where he grew up, noting its distinctive flora and fauna. His father was unimpressed, still hoping his son, the second of four children, would be able to take over the family business one day. It took an uncle to appreciate his curiosity, so unusual in his small provincial town, and beg his parents to send him to Paris for school. They only relented when he passed the rigorous examination to the École des Mines, or School of Mines, in Paris. The mining business was booming as coal and its extraction became more important than ever. The program in Paris was the best of its kind, with its graduates sought after to run mines across the globe.

By the time he had graduated in 1828, both of his parents had passed away, his mother first in 1826 shortly after he had left home, and his father in 1828 while he was touring mines in Germany. With no more parental expectations to uphold and a newfound love of chemistry, Laurent had entered Dumas's lab to start again at the bottom. But his mining background proved surprisingly useful when their work took an unexpected turn. Dumas and Laurent had continued working together on camphor while also extending their research to include the essential oils of peppermint and anise. The cholera scare had driven demand for peppermint so high, in fact, that France began importing it from North America for the first time, and Dumas noticed that this particular variety could, like anise, be transformed into a camphorlike solid concrete at low temperatures. He ran these substances through the kaliapparat and determined their formulas. Remarkably, they all had equal amounts of carbon and oxygen, and their ratios of hydrogen fell into a satisfying series. For every ten volumes of carbon, Dumas figured, peppermint concrete had ten volumes of hydrogen, camphor had eight, and anise concrete had six.[49] Was there a substance that completed this series, with four volumes of hydrogen, and the same amount of carbon and oxygen? There was: a coal by-product, naphthalene.

The chemists of France had virtually no experience working with coal, a substance they left entirely to the mineralogists. Taxonomies had always classed it among the minerals, with Linnaeus placing it in the *Regnum Lapideum*, which included minerals, rocks, and fossils.[50] It was, in fact, technically a fossil itself since the original meaning of the term simply meant something dug up out of the ground.[51] It had only been a few years earlier that it had occurred to anyone that the coal was once alive. Dumas's father-in-law Alexandre Brongniart, a graduate of the School of Mines trained to hunt for coal, had agreed in 1808 to help Georges Cuvier map the geological strata underneath Paris. Cuvier had been the first to establish that most fossils were the remains of long-dead extinct species, and it dawned on Brongniart that coal, which always appeared in layers together with fossilized plants, might also be a remnant of these ancient worlds.[52] Dumas followed the line of thought, using the category of "fossil combustibles" to describe coal and its products.[53] He acknowledged that these were plants that had undergone a "long and profound alteration," but he still felt that geologists, not chemists, had the most useful language for them, and that the categories based on the epoch of formation—lignite, coal, and anthracite—should stand.

English chemists were less squeamish about working with coal, and it was an Oxford professor, John Kidd, who first made the substance of naphthalene, which he named after the *naphtha* that ancient Greeks had used for the crude oil they occasionally found oozing from the ground.[54]

FIGURE 22. The gasworks at La Villette was one of a few factories in Paris that made coal gas for illumination. Coal tar was a by-product of the process, which they disposed of or sold very cheap.

He got it by distilling coal tar—itself the product of a recently invented procedure that heated coal to extraordinary temperatures in an airless kiln. This "destructive distillation," as it was called, gave three different products, the most prized of which was a solid, charcoal-like block of nearly pure carbon left behind in the kiln. Known as coke, it burned hotter and cleaner than normal coal, and was the best form for smelting iron. The process also made a gas called, with little imagination, "coal gas," which could be burned as an illuminant in gas lamps. Last and definitely least was the greenish-brown liquid called "coal tar." It was mostly a waste product, although the British found some use for it as a topical medicine to treat skin conditions and preserve railroad ties. Michael Faraday, director of the Laboratory of the Royal Institution, was the first to give its chemical formula, in 1826.[55]

No one in France had ever made naphthalene. It was a filthy, messy business, and Dumas assigned the entire operation to Laurent. The first step was to procure some coal tar. Fifteen years earlier that would have been a near-impossible proposition, but now it was easy. Paris had begun switching over to gas lighting in the 1820s, replacing its old whale-oil lamps with the cheaper, brighter coal gas. There were now several gas works in the city, and they generated a fair amount of coal tar as a by-product which collected, unwanted, in underground cisterns. It was not hard to get a little to bring back to the lab. Turning it into naphthalene was another story. Straightforward distillation gave only inconsistent results, and Laurent cycled through a number of different techniques.[56] The best results came when he put the coal tar in a retort and passed a current of chlorine over it for four days. It was not a pleasant process to be around. As he heated up the resulting liqueur, it gave off vapors of hydrochloric acid "of a disagreeable odor" that billowed throughout the courtyard of the École Centrale, choking and gagging anyone unlucky enough to get in the way.

What was naphthalene? It was a white solid, like camphor, but smelled even stronger, somehow taking camphor's sharp pungency and amplifying it. It had "such a strong and penetrating odor," Laurent warned, "that it was enough to touch it with the tips of one's fingers to be impregnated with it for 4 or 5 days." It was not entirely objectionable, with its clean, medicinal tones, but it was not exactly pleasant, either. We would recognize the smell immediately as mothballs, which came to be its standard use in the twentieth century. But at the time it was hard to think of a commercial use for something whose chief attribute was its bracing and aggressive smell. This was unfortunate for Laurent because he had figured out how to make a lot of it very cheaply, pointing out "that it would be possible to deliver it at low price to commerce, if one could use it."[57]

Laurent noticed something else about naphthalene that sent him in a direction orthogonal to Dumas's original instructions. It had, he said, a "great tendency" to crystallize into thin sheets. This was not unusual—many compounds formed crystals under the right conditions. But Laurent

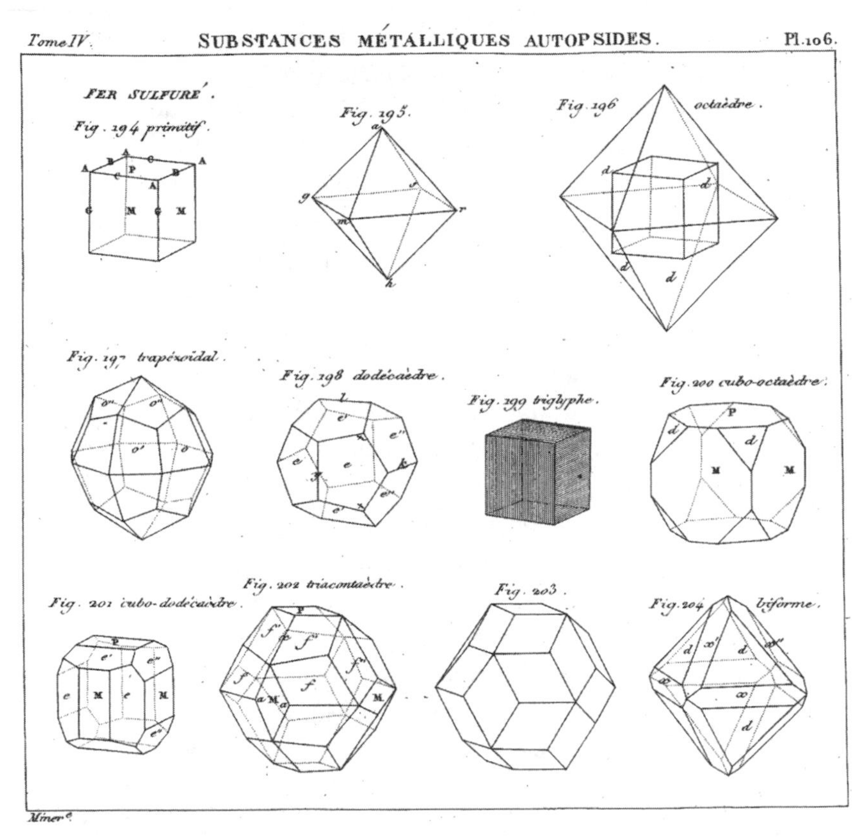

FIGURE 23. From René Just Haüy's textbook, *Traité de mineralogie*. He described each crystal in terms of its "fundamental nucleus" or "core molecule."

paid closer attention than other chemists. Crystals had been important in the early days of alchemy, sought after in the same way as gemstones and precious metals. But by the nineteenth century, their study was largely confined to a new field—mineralogy—that ran parallel to but separate from chemistry. Its founder, René Just Haüy, was born the same year as Lavoisier, served with him on the Commission on Weights and Measures, and was to the field what Lavoisier had been to chemistry, giving it its first rigorous basis. He taught at the School of Mines and curated its cabinet of mineralogy. Although he was no longer there by the time Laurent entered the school, his techniques remained central to the education of the mining engineer.

Haüy's chief insight was that every crystal had a "fundamental nucleus," or smallest possible shape into which it could be broken down. The rumor was that he first noticed this after dropping a priceless crystal from the Museum of Natural History onto the floor, and all the broken pieces repeated the same shape. Haüy had identified six different polyhedra that served as the "core molecules" (*molecules integrantes*) or "nucleus" of crystals, and these aggregated in repeating patterns to give crystals their structure. He made wooden models for his students, which they fit together like a puzzle to see what kind of crystalline structures they could make. Haüy also developed an instrument called the goniometer, which could measure the angle between crystal faces and identify this internal structure. Every student at the School of Mines learned how to use one, and Laurent brought his with him to the École Centrale.

Laurent had always been an artistic sort, constantly doodling in the margins of his scientific notes, and he had developed a knack for picturing the three-dimensional patterns in which crystals fit together. As he measured the angles of the naphthalene crystals, he began to wonder if they might offer a clue to the question that Dumas and Biot had recently been fighting over: was there any way to know what the internal organization of a chemical molecule looked like? Perhaps, thought Laurent, the unit structures that Haüy had seen in crystals actually reflected an even more fundamental unit structure of the molecule itself. Laurent became convinced that the properties of organic molecules could be explained by their internal structure, and wanted to use Haüy's classification of crystalline forms as a model for how to do it.[58]

He started with the question of why some chemicals reacted with others. The reigning theory of the day attributed it to the interaction of positive and negative charges. The prominent Swedish chemist Berzelius had named this "electrochemical dualism," dividing all molecules into an electrically positive part and an electrically negative part. Dumas elaborated on this model with his "theory of substitutions," explaining certain reactions as a swapping out of these electropositive and electronegative components.

Laurent hoped that his ideas on the internal structure of the compound could provide a theoretical underpinning to explain Dumas's substitution theory.[59] Perhaps the stable, electropositive part functioned as a nucleus, much like Haüy's crystalline nuclei. Starting with naphthalene, he began substituting various components for one another. But his results did not support Dumas at all. Instead, they convinced him that the theory was entirely unworkable. For one thing, he found that he could substitute chlorine for hydrogen, and the properties of the original compound and the substituted compound were not that different. For Laurent, this was an indication that what really mattered was the position they occupied in the molecule's structure. But it was a disaster for electrochemical dualism, since chlorine was electronegative and hydrogen was electropositive, and they should not have been interchangeable. His work came across as an attack, not only on Dumas but on the entire received wisdom of European chemistry.[60]

Berzelius and Liebig wrote to Dumas from abroad, warning him that his assistant was out of line. Dumas turned on him immediately, publicly disavowing what he called "the inexact experiments of M. Laurent."[61] It was the worst thing you could say about a chemist, particularly for Laurent, who took pride in his experimental precision. As the relationship between them disintegrated, Dumas found a way to get Laurent out of his lab. His father-in-law, Brongniart, needed a new director of chemical analysis at Sèvres, and Dumas urged Laurent to go. Laurent had no interest, bothered that he would have to live in Sèvres, far enough outside of Paris to cut him off from the scientific community. But Dumas made it clear that this was not a request but a demand, and if Laurent wanted a future in chemistry, he had better go. As Laurent put it, "I left for Sèvres well aware that the game was up for me."[62]

The job turned out to be just as bad as he had feared, He worked mostly with hydrofluoric acid, used for etching porcelain. It was an insidiously nasty substance, able to penetrate the skin without a surface burn and circulate, unnoticed, in the body, where it wreaked havoc—destroying lungs, corroding bone, stopping the heart. Brongniart had promised that the job would be a stepping-stone to greater things, but those never materialized. Laurent left as soon as he could, alienating even more people in the process. He had no other position awaiting him, no powerful mentors or benefactors on his side. But he had his chemical equipment and a head full of ideas, and he was eager to try to make it on his own.

8

Radicals and Bohemians

PARIS, COUR DU COMMERCE, 1834

Laurent was, he acknowledged, "left without resources, without friends, without protectors." But if you were going to be young and poor in the 1830s, with nothing but your own burning ambition, then the Left Bank of Paris was not a bad place to be. It had always been the shabbier side of the Seine, its streets less paved, its fashions less refined. But by the 1830s, it had taken on a bohemian élan. Poets, painters, writers, and musicians joined the students and scholars who had long claimed the left side of the river, drawn by the cheap rents and many schools. Laurent joined them. "I threw myself into science to find there some distraction," he wrote, following his own particular muse.[1]

Laurent rented a tiny garret apartment in the cour du Commerce, off the rue Saint-André-des-Arts. It was a narrow, lightless courtyard, hardly more than an alley, and known for its cheap student lodgings. An inauspicious door on its side was the back entrance to Le Procope, the oldest café in Paris. Long a den of radical thought for the likes of Voltaire and Marat, by the 1830s the café was a center of socialist and literary circles, a favorite of George Sand, Alfred de Musset, Victor Hugo, and Honoré de Balzac. By this time, Balzac had given up his fashion columns to become something of a bard to the neighborhood's offbeat strivers. In *The Atheist's Mass,* he made two of his most impoverished characters wander the streets of Paris looking for a room, pulling everything they owned behind them on a cart. They stopped at each place that had a notice of rooms to rent, only to find, time after time, that they could not afford the price asked. Only as the day drew to a close did they find a place cheap enough: the cour du Commerce, where they settled, like Laurent, under the slanted roof of an attic room.[2] Laurent's neighbors included the Romantic poet Charles Augustin Sainte-Beuve, who was currently caught up in a bit of a scandal. He had been using his small apartment, next to Laurent's, to conduct an affair with the wife of his close friend and mentor, Victor Hugo, and it was all spilling out in the public.

Laurent struck up a friendship with Édouard Laugier, who was also struggling as an outsider after losing his position teaching chemistry. The two collaborated on various projects, and Laurent also tried to support himself by following Édouard's model of teaching chemistry classes based on practical techniques to supplement university lectures. Where Édouard had concentrated on the students studying chemistry at the Collège de France and the Sorbonne, Laurent targeted the students at the nearby School of Medicine. The courses went well, and he never lacked for students. But he resented the time they took away from his own research and tended to take a break whenever he had made just enough money to survive. One of his brothers, visiting from home, was shocked when they sat down to dinner with nothing on the table but bread, cheese, and "badly filtered water" still cloudy with sediment.[3] But you did not need money to enjoy life in the Latin Quarter of Paris. Laurent was a gifted musician and a talented artist, the true currency of style in the salons of the Left Bank.

Laurent soon fell in with a group of radical socialists, introduced by an old friend, Jean Reynaud. The two had been students together at the School of Mines, entering and graduating the same year.[4] Reynaud had then gone to Corsica to work as a mining engineer and make a geological map of the island. But, inspired by the revolution of 1830, he had returned to Paris and joined a newly formed utopian socialist movement, the Church of Saint-Simonianism.[5] By the time Laurent moved into the neighborhood, Reynaud had left the church, but he held onto its vision that technology could bring about a new era of equality and mutual understanding. He had joined forces with Pierre Leroux, a philosopher who had just introduced the word "socialism" into French political discourse in 1834, and this novel concept drove their work.

Reynaud and Leroux invited Laurent to join them in their latest project: a socialist revision of Diderot's *Encyclopédie* that would not only incorporate all of the new technology rapidly accumulating in the world, but also do it in a way that would contribute to the liberation of the working classes. They called their project the *Encyclopédie nouvelle,* and Laurent was soon part of the inner circle, contributing several articles, including Chemistry, Combustion, Cobalt, and more.[6] His work hewed to the project's bigger themes, highlighting the practical benefits that advances in chemistry would bring to the people. But he also took the opportunity to explain his love for chemistry, and its ability to reveal aspects of nature normally hidden from sight. It was always astounding, he wrote in his article titled "Chimie," to see the way it could "expose the intimate properties of nature, explain all of its mysteries."[7]

But not quite *all* of its mysteries. His article on fermentation showed the limits of chemical explanation. Fermentation was a puzzle yet unsolved, now widely acknowledged to be more than a straightforward chemical reaction. Somehow, Laurent wrote, the organic matter was able to undergo profound transformations, even after it was dead and cut off from the "vital force" that had animated it during life. He began by posing questions: "How does it differ from ordinary chemical phenomena? What does it present of the extraordinary and inexplicable?"[8] Laurent offered his best explanation—a rather strange, idiosyncratic view of fermentation as a form of decomposition in which plant substances lost their internal organization. Details were fuzzy, but the article reveals how Laurent was thinking about the chemistry of living things. He drew parallels, in the article, with the organization of crystals and mentioned that he was experimenting on naphthalene and the essence of bitter almonds to further explore the matter.

These experiments were the same ones he had been performing with Édouard. Still smarting over Dumas's comments and determined to overthrow the electrochemical theory, Laurent had continued swapping various substances in and out of naphthalene.[9] Chlorine, bromine, nitric acid—he busied himself with combining ingredients, heating them, cooling them, stirring, evaporating. His efforts produced thick, red vapors, an oily substance never seen before.[10] One of his new creations had such a tenacious, unbearable odor that, he said, "after handling it I could not, even three or four days after, present myself in a public space without provoking loud complaints, even though I had taken baths and completely changed my clothes."[11]

Not everything he made smelled bad. At one point in his series of reactions, he produced a secondary oil that gave off the smell of bitter almonds after he left it out in the air for several days. He had wanted to do more experiments to follow through, but noted in his paper that he had run out of the small amount of materials he had access to.[12] He had, by this time, dismissed his paying students to give himself more time for his research, and there was no money coming in. Édouard, however, had access to sacks of bitter almonds in his family's storerooms, and the two teamed up to continue the investigations.

Laurent and Laugier got to work. Across the cour du Commerce, at the Café Procope, Balzac also found himself in a flurry of activity. In two furious months over the summer of 1834, he dashed out a new novel, *La Recherche de l'absolu,* whose plot had occurred to him a few weeks earlier. It followed the tragic path of Balthazar Claes, a wealthy man who drove himself to poverty and isolation with his obsession to discover the secret to the chemistry of life. He locked himself away in his laboratory, spending freely on expensive reagents, while his fortune and neglected family withered away. The book went into print by October 1834, and Balzac inscribed a dedication on the title page of the first edition: "To M. Laugier, in token of gratitude from the author, who was not much of a *chemist.*"[13] Literary circles have debated just who this Laugier was, who coached Balzac in chemistry for the book. Thoughts went first to André Laugier, with the drawback that he had died several years earlier. Speculation then centered on his son, Ernest, who was, however, an astronomer, and did not seem to know Balzac. More likely is that it was Édouard, a chemist then in search of his own absolute. He was at that moment gaining a reputation as the master of alchemy's most cherished form of manipulation: distillation.

PARIS, PLACE DE LA CONCORDE, 1834

Across the Seine on the Right Bank, four large pavilions had sprouted on the Place de la Concorde, filling the empty space that had a generation ago housed the scaffold of the guillotine. But these were monuments to a different kind of revolution, an industrial one that would bring about a new era with, one hoped, fewer heads lost. It was the latest Exhibition of Industry, a tradition that had begun as the brainchild of Chaptal and Napoleon. The first exhibition, held in 1798, had drawn 110 participants. This one, the eighth, now drew 2,447 exhibitors vying for space in the cavernous halls. Originally scheduled for the spring of 1832, it was postponed because of the cholera epidemic, and by the time it opened in 1834, Parisians were desperate for a glimpse of the latest innovations.[14]

The Central Jury of the exhibition, headed by Thénard, opted for a novel organizational scheme, arranging things according to their "social goals." Laugier Père et Fils thus found themselves in the gallery of "sensory arts," defined as those goods having the end goal of "pleasing the senses of man."[15] It turned out to be a particularly pleasant corner of the exhibition, with winemakers offering samples and extravagant displays of artificial flowers. There were all the latest innovations in paints, wax, candles, wallpaper, and corsets, along with genuine novelties like foldable umbrellas and self-administering clysters "of varying elasticities."[16] The biggest draw could be found in the carpet section: a rug made from stitching together the skins of hundreds of cats, on sale for 10,000 francs.[17]

The jury gave its top mark in the soap and perfume category to Laugier Père et Fils and awarded a silver medal, the highest honor of the section. They noted that Laugier the father had a "colossal reputation" among perfumers, and Laugier the son, "already known as a chemist,

FIGURE 24. The 1834 Exhibition of Industry, at the Place de la Concorde. The obelisk in the center, from Luxor, Egypt, was delayed in transport and did not arrive until after the exhibition began.

having deep knowledge of a difficult science, can only add to this reputation."[18] They had particular praise for how Édouard, "enlightened by theory," had used chemical principles to make his soft, transparent soap.[19]

His novel soap had won him a silver medal, but his primary enterprise remained the art of distillation. He was at that moment developing a new distilling apparatus—a modern alembic specially designed for the production of *esprit de vin* and *esprit recteur*. Its signature piece was what Édouard called the "dephlegmator," where the final battle took place in the distiller's war between spirit and phlegm. The language of distilling was still, at heart, alchemical, with "spirit" as the volatile quarry and "phlegm" the nasty-smelling remnant that befouled it, described in chemistry textbooks as "the odor of the breath of drunk people, when they are badly digesting the wine."[20] The goal was to get as much phlegm out as possible, and Édouard would soon be better at doing this than anyone else in Paris.

The dephlegmator was a tall vertical column divided into a number of different levels by a series of "trays." The trays served as steps on a ladder, with the ethereal vapors climbing upward and the less volatile fluids flowing down. Each tray had its own equilibrium, with the vapors becoming richer and richer in spirit as they rose. In this sense, the dephlegmator was a kind of helical purgatory, where at each stage the soul of the distillate was weighed and judged. Too heavy and laden with foul-smelling phlegm, and it was cast downward into the heat of the boilers. Sufficiently diaphanous and pure, and it could ascend upward to the cool tunnel of the condenser, where a transformed state of existence awaited.

In a standard alembic, the distillate at the end would only be partially "dephlegmated," so the distiller would usually pass it through the alembic again once or twice, a process known as "rectification." Édouard turned this into a process of "endless rectification," where, instead of removing the distillate at the end, he rerouted it so that it would trickle back down the column, joining the rising vapors in the ongoing process of weighing and judging.[21] Previously, three passes through a still was considered high quality, but for Laugier, this number approached infinity, an unending cycle of rising vapors and falling waters that did not stop until it could not be purified anymore. Some twenty years later, the French journal *L'Illustration* would look back on the moment as an apotheosis of the distiller's long-held dream, noting that Édouard Laugier had so perfected the process of distilling that "it seems impossible that this art undergoes any new modifications."[22] Thirty years later, Nicolas Basset deemed Édouard's apparatus the best of its kind in his textbook on alcohol distilling, remarking that its only drawback was that there was no way to cut the process short if you were willing to settle for an inferior product.[23] But all of this praise was far in the future. In 1834, the only people to appreciate his distilling techniques were the customers of Laugier Père et Fils and his collaborator, Laurent.

By this point, Laurent had joined Édouard as an employee of the shop on rue Bourg l'Abbé, with both of them entering into an arrangement for a share of the profits for the time that they worked. Being used to

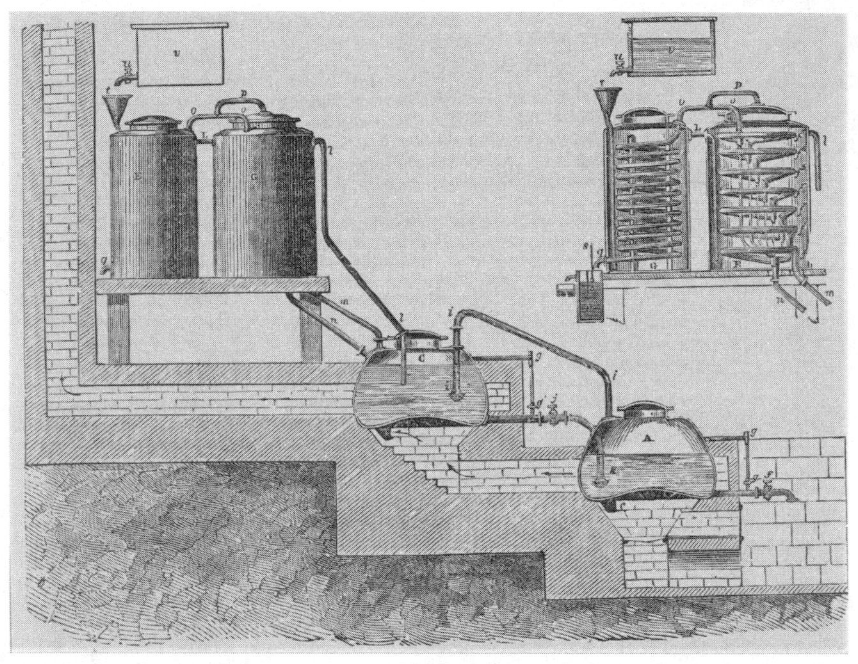

FIGURE 25. Édouard Laugier's distilling apparatus. It consisted of a boiler (A), a second boiler or exhauster (A'), a dephlegmator (B), and a condenser (C). The reservoir (D) contained the fermented must (typically wine for Laugier). The liquid went into the funnel (E), eventually making its way through tube F into the first boiler (A) and, if the valve (R) was open, into the second boiler (A'). As the boilers evaporated the liquid, the vapor would rise through the tubes T and G into the dephlegmator. There, the delicate separation took place as the vapor rose through seven helical sections. The watery part (or phlegm) condensed out and ran back into the boilers through tube H. The purer distillate, meanwhile, continued its rise and then passed into the serpentine condenser (C), eventually emerging in liquid form from the spigot (N).

poverty, Laurent did not even demand a salary but took only five francs now and then to procure his most urgent necessities.

GIESSEN, 1832

Meanwhile, farther north, Justus Liebig's institute for practical chemistry was thriving.[24] Although the university still would not support it, he had built a new laboratory himself on some adjacent property, in the guardhouse of an abandoned barracks. He lived in an apartment above, together with his growing family. Although Liebig's relationships had tended to be with men when he was a student, he married a woman, Henriette, in 1826, and they soon had a full household of children.[25] But he spent most of his time in the laboratory downstairs, which was open for students from early morning until well into the night. They were a particularly devoted group and had started wearing buttons on their jackets to distinguish themselves, fashioned in the shape of Liebig's recent invention, the kaliapparat.[26] This device, and its ingenious method of counting up carbon, had given the Giessen laboratory an international reputation for the analysis of organic products, and it was often visited by foreign chemists. One of these, in 1832, was Jules Pelouze, another of Thénard's assistants at the Collége de France, who provided Liebig with all the latest news from Paris, in particular Pierre-Jean Robiquet's latest breakthrough in his study of bitter almonds.[27]

Robiquet's obsession with the smell of bitter almonds was now stretching into its third decade. "Organic products are so complicated," he pointed out, "that it is only by force of returning to them again and again that one is able to untangle their true composition." He had remained puzzled as to why two extremely volatile substances like cyanide and the essential oil of bitter almonds could remain inert for so long in almonds, even as they were being ground into a paste. His latest step was to separate out "a particular crystalline substance" that he named *amygdalin*, from the Latin word for almond.[28] Robiquet was convinced that amygdalin could be "activated" by reacting it with some other substance, and it would then transform into cyanide and the fragrant essential oil. But nothing he tried worked. (We now know that amygdalin is broken down by an enzyme that almonds store in a separate location, but there was no concept of enzymes at the time.) He did, however, produce something profoundly unexpected: a sweet-smelling lump of white crystals ultimately identified as benzoic acid.

Benzoic acid was a curious substance, first discovered it in the sixteenth century by the alchemist and prophet Nostradamus. He had heated the aromatic resin gum benzoin over a "great fire as strong as possible" and found that it sublimated into white crystalline flakes that he described as "snow." Alchemists would come to call the substance "flowers of benzoin."[29] Lavoisier renamed it the substantially less poetic benzoic acid. But it remained a staple of both pharmacists and perfumers, its pleasant, vanilla-like taste being used, for example, to mask the bitterness of opium in the paregoric elixirs popular in Britain. And, if anything, its nature became more mysterious. Up until that point, gum benzoin had been the only source of benzoic acid. But chemists soon began to find it in the strangest of places. It first showed up in the balsams of Tolu and Peru, which was perhaps not so strange since they were also both fragrant tree resins with a similar olfactory palette. But then it appeared in the urine of newborn babies, in castoreum, the secretion of the beaver's anal scent gland, and in a mysterious substance that smelled like cow dung brought back from a cave on the Isle of Capri.[30] So when Robiquet found it in the products of amygdalin, it raised even more questions, and he noted that "this singular product, which has already been examined so many times, is still far from being known."[31]

Pelouze had brought some of Robiquet's amygdalin with him when he arrived, and Liebig was eager to begin working with it. But just then he received a letter from Friedrich Wöhler, who was teaching in Kassel. Liebig and Wöhler had remained friends after their exchange over isomers. Wöhler had also settled down, marrying a cousin, Franziska, and starting a family. But his letter now brought terrible news: Franziska had died giving birth to their second child.[32] The child, a daughter, had survived, but Liebig nonetheless thought the best course of action was for Wöhler to leave the scene of tragedy and come join him in Giessen, where they would make their sorrow "disappear in work." He wrote back on 15 June 1832, begging him to come and enticing him with the promise of "some amygdalin from Paris" and "25 pounds of bitter almonds to work on."[33] Wöhler arrived in Giessen soon after, and the two got to work.

Endless hours of distillation reduced their pantry full of bitter almonds to a few precious vials of essential oil. They then proceeded to react their pure oil with everything they could throw at it. They exposed it to chlorine gas, dissolved it in nitric acid, attacked it with fluorine, bromine, ammonia, and more. These reactions produced new products, which Liebig and Wöhler in turn reacted with more things to produce a wider and wider array of new substances never seen in nature. They were looking for any hint of something, anything, that stayed constant through all these various transformations. They carefully analyzed all their products, counting up the numbers of carbon, hydrogen, and oxygen in each. Then they balanced the ledger books of the chemical reactions, making sure the numbers in the "before" column matched the ones in the "after" column. Across all the reactions and the multitude of different products, they kept seeing the same cluster pop up again and again: fourteen carbons, ten hydrogens, and two oxygens.

This was it: the one constant in the chaos. The "single compound," they said, "which preserves its nature and composition unchanged in nearly all its associations with other bodies."[34] They called it the "benzoyl radical." The term "radical" had been part of Lavoisier's theory of acids—they were the "root part" of the acids onto which oxygen could attach or detach itself. The radicals that Lavoisier worked with in the inorganic realm were pretty simple. For example, the radical of nitrous acid was nitrogen, the radical of phosphoric acid was phosphorus, and so on. Liebig and Wöhler hoped to adapt the theory to the more complicated world of organic chemistry, where the radical would be a clustered group of carbon and hydrogen. But it would also function as a stable "root part" through its various reactions, thus making the life of organic chemists much easier. They would no longer have to wade through a swamp of undifferentiated carbon but could organize them together in neat groupings. Berzelius, when he heard the news in Sweden, welcomed it as "the beginning of a new day in vegetable chemistry" and suggested they rename the benzoyl radical *proin,* from the Greek word for "break of day," or *orthrin,* meaning "dawn of the morning,"[35] Liebig and Wöhler kept the original name but agreed that it marked a new day for organic chemistry, admitting that "we have reason to congratulate ourselves" for finding "a point of light" in "the dark province of organic nature."[36] As Liebig wrote to Pelouze in 1834, "with the impetus that organic chemistry has acquired, one truly does not know where it will stop, and where we shall be led. One becomes dizzy from so many discoveries."[37]

Liebig and Wöhler published their results in a new journal that Liebig had just started, the *Annalen der Pharmacie.* The journal would, he hoped, make him a bit of money, but more than that, it would be a place for him and his students to publish their work, and it would serve as the vehicle for a new research school. It was not long before he was using it to pick fights with his rivals. One of his most keenly felt rivalries was with the chemist Eilhard Mitscherlich, who now held the plum job of professor of chemistry at the University of Berlin. Liebig had watched his rise with some distrust, admitting to Berzelius in confidence that he thought Mitscherlich "would not hesitate for an instant to disown the ladder that raised him if it were to his advantage."[38] But the two had maintained outwardly friendly relations. Mitscherlich even visited Liebig in Giessen in the spring of 1832, where he admitted his rival was "amiable, clever, and shy," noting that others surely found that appealing, and then Liebig returned the visit in Berlin later that fall.[39]

Mitscherlich had started his own experiments with benzoic acid and had discovered a different substance that he thought was the key to everything. It was a colorless liquid with an inoffensive, sweet smell that he had found by distilling benzoic acid in the presence of slaked lime. He named it benzin (or benzene in English), and determined it had equal volumes of carbon and hydrogen. Proposing that it functioned as a "chemical unit," Mitscherlich used benzene to make a string of different kinds of acids and wrote a paper, titled "Regarding Benzene and Acids of the Oil and Tallow Variety," which he sent to Liebig to publish in the *Annalen.*[40]

Liebig was having none of it. He published the paper, but with the unusual step of adding a nine-page "addendum from the editors" refuting its main points. He insisted that benzoyl was the only possible radical and that it was "self-evident" that benzene was nothing more than a decomposition product, a stray bit of leftover detritus.[41] Faraday, he pointed out, had found the same piece of detritus and called it bicarburet of hydrogen. His final sentence was a masterpiece of passive aggression. "In a case like the present one, where intimations, speculations, and conclusions may mislead the reader into time-wasting and pointless chases after organic radicals, I believe it necessary to express my own opinions in all candor and irrespective of the friendship that binds us."[42] In private, Liebig admitted he was glad to be done with the "half-cursed friendship."[43] He felt nothing but relief, he wrote to Wöhler, that he had poured out all his bile and transformed any pretense "into a clear, open hostility."[44] Wöhler begged him to reconsider, saying that he already had a reputation in both France and Germany as "someone who likes to fight," and that his attacks only reflected badly on him.[45]

Liebig's attack kept everyone focused on the benzoyl radical as the most promising path forward in organic chemistry. The benzoyl radical, remaining constant through a kaleidoscope of reactions, could tie together a world of chemical substances whose connections seemed random and puzzling, from bitter almonds and peach pits to fragrant tree resins, beaver secretions, and baby urine. The only problem was that it was still something of a fiction. No one had yet isolated the substance. It was not for lack of trying. Robiquet, when he heard the news, immediately set about trying to isolate the radical, renewing his work on bitter almonds once again. Laurent and Laugier did the same, racing against the most established names in chemistry. But it remained stubbornly elusive.

PARIS, RUE BOURG-L'ABBÉ, 1835

Laurent cut a rather strange figure at the upscale perfume shop, where he would occasionally station himself behind the counter, "making his observations on the vanities and prejudices of the elegant and high-bred persons who frequented it."[46] He had abandoned neither his radical principles nor his Bohemian appearance. All around him, however, fashion was reaching the high point of an exuberant Romantic phase. Leaving behind the simple austerity of Empire dresses, women now went out adorned in layered petticoats and flowing ribbons, the sleeves on their dresses so voluminous they used metal supports to keep them puffed in a mutton-chop shape. Hats and bonnets had never been larger. Men, too, strove for a dashing flourish in this age of the dandy, with their padded shoulders and cravats arranged in a dramatic "waterfall" style to enhance the volume of their chests. Perfume was a necessary accompaniment to the extravagance in style, and sweet, floral scents had never been so popular. The bracing astringency of an eau de cologne's citrus notes gave way to an unashamed sweetness, with soft, redolent fragrances recalling a bouquet of flowers. Roses were the queen of the moment, with violets and lavender attending. And more shelf space than ever was devoted to soap, as the habit of bathing had begun to take hold widely. Rose soap was a big seller, along with another new favorite: almond soap.

Almond soap soared in popularity in the 1830s, at least in part because many people confused it with the expensive *pâte d'amandes,* which Marie Antoinette had used to preserve her youthful complexion.[47] Modern almond soap, however, was not made from ground almonds. It was just normal soap, made from animal fats and scented with the essential oil of bitter almonds, which chemists had so recently learned to separate from cyanide. Laugier Père et Fils sold it all: the soap, essence, and *pâte* alike. Bags of almonds filled its storerooms, and as news of Liebig and Wöhler's discovery came to France, Laurent and Édouard redoubled their investigation of the enigmatic bitter almond.

But it hid its secrets well. "The essence of bitter almonds is the Proteus of chemistry," complained Laurent, referencing the prophetic sea-God of Greek mythology who knew all things, past, present, and future, but constantly evaded efforts to pry this knowledge from him. When cornered for questioning, he transformed into some other being—a seal, leopard, serpent, or water—to slip away unapprehended. So too, said

Laurent, did the products of bitter almonds shape-shift unexpectedly, with the same set of operations producing maddeningly different results. It made a difference whether the water was from a river or a well, and whether it had traces of limestone or carbon dioxide in it. It made a difference whether they distilled it with the fire on the bottom (*per ascensum*, he said, using the alchemist's phrase) or on the top (*per descensum*).[48] Most important, Laurent pointed out, it was hopeless to use commercial preparations of the oil, as the expensive product was so often adulterated with peach pits, apricots, or sweet almonds that threw off the results.

Laurent and Édouard pursued their elusive quarry with the wiles of Homeric heroes. Édouard was in charge of distilling. He built "a new, very ingenious apparatus that he just set up," as Laurent described it, that allowed them to pull out samples to investigate at various stages of distillation.[49] It still took many repeated efforts, varying the conditions ever so slightly, before Édouard found his novel resinoid substance in the bottom of this flask and called in Laurent to investigate.[50] The resinoid appeared to contain several different substances within it, but they were able to separate them by dissolving and filtering it several times. Laurent took one of the substances, passed a current of chlorine over it (the technique he had perfected with naphthalene), and then dissolved the resultant product in alcohol and crystallized it. The formula matched the long-sought benzoyl radical. He carefully measured the angles of its "beautiful prisms," observed how they broke, and then placed some in his mouth to see how they felt ("a disagreeable sensation" was the verdict).[51] They had, they believed, captured Proteus and could now demand its secrets.

There was another meaning of Proteus, too. Alchemists had adopted the term long ago to mean a "First Matter" capable of becoming all things animal, vegetable, and mineral, an essential part of the search for the Philosopher's Stone.[52] And although alchemy had fallen into disrepute, chemists of the 1830s pursued a similar goal—seeking a fundamental, protean root behind the sprawling diversity of organic molecules. Laurent and Édouard continued their work with bitter almonds to see what other substances they could find (Laurent published under his name alone, but he thanked "my friend, Ed. Laugier" for providing him with his materials.[53]) They pulled out samples of every product they found in the course of its distillation. Laurent treated the samples with ammoniac and got a mix of various substances that would have been impossible to analyze with traditional means. But Laurent had a trick up his sleeve, thanks to his hours in mineralogy class learning how to measure crystals. He crystallized the materials and looked for any differences. Using the most powerful microscopes, of at least 300×, he was able to sort the different substances based on their shapes, and then analyze their compositions.[54]

He soon found the same product that Mitscherlich had three years earlier, that had precipitated his fight with Liebig.[55] Mitscherlich had called it "benzin," highlighting its origins in benzoic acid, and, ultimately, bitter almonds. Faraday had called the same substance "bicarburet of hydrogen" in 1825, after deriving it from coal tar. Its presence in coal tar had led Liebig to condemn the substance as a bit of incidental rubble, a dead fragment that could not possibly be important in the processes of life. But Laurent was not so sure. He had already begun to suspect an important connection between the two, and that the by-products of coal tar were not as dead as they seemed. He gave the substance a new name of his own devising: phène, from the Greek word *φαινω*, meaning "I light." It was a reference to the gas now illuminating street lights throughout Paris, emphasizing its origins in coal tar.[56]

The discovery rekindled his ambitions to explore the connections between bitter almonds and coal tar. He thought back to his accidental production of the oil that smelled like bitter almonds, and the foul-smelling substance that had preceded it, so pungent that people complained how badly he smelled for several days afterwards, no matter how many baths he took. He had never published his results when he first got them in 1832, but he went back to them in 1836, naming the malodorous substance "acide chlorophénisique," based on his new concept of *phène*, "I should have wanted to take the subject back up," he admitted, "but it required me to once more plunge myself into unbearable odors; and besides, I no longer have a suitable space for such experiments." Four years ago, he pointed out, "I had no neighbors to bother," but now he was doing his work in the back of perfume shop, and filling the air with unbearable stenches was not an option.[57]

He switched instead to working with the essential oils of things like cinnamon, tarragon, and savin juniper, as well as plants such as indigo and camphor. But at the heart of it was the same insight that he had reached when working with naphthalene: that organic molecules had a spatial arrangement in the same way that crystals did. Just as Haüy had reorganized mineralogy around the geometrical shapes that were the most fundamental units of crystals, so too did Laurent propose to reorganize organic chemistry around the fundamental unit he called a "nucleus." He envisioned this nucleus as a collection of hydrogen and carbon which, like a crystal, had the form of a regular geometrical shape—a pyramid, square, hexagon, etc.—with the carbons at the angles and the hydrogens on the edges. These hydrogens could be replaced by chlorine, or other elements, to give another substance, which Laurent called a "derived nucleus," but insisted the prism would retain its overall shape and have very similar properties to the original or "fundamental" nucleus.[58] The fundamental and derived nuclei formed together what Laurent called a "series," such as the benzoic series based upon the benzoyl radical. Camphor, cinnamon, and naphthalene had their own series. The key, for him, was that each series was governed by a hydrocarbon that served as its nucleus. The relative ratios of carbon and hydrogen determined the particular hydrocarbon governing the series. The nucleus then combined with various elements to form the molecules that made up a series, with particular consideration given to their form and arrangement.

It was a monstrously ambitious plan. He wanted nothing less than to make a clean sweep of all the cluttered, confusing relationships currently in place, and provide a new classification based on clear, rational principles.[59] His focus on the shape of the molecule placed him at odds with the other chemists of his day. He did try to reconcile with their approach, sometimes referring to the "fundamental nucleus" as the "fundamental radical," and the "derived nucleus" as the "derived radical." But he meant something very different than Liebig had. Liebig had treated his radical as a compound substance that behaved like a single element, always remaining constant through reactions. His primary focus was the chemical formula, and the role of the radical in balancing the ledger books when counting up the carbons and hydrogens. For Laurent, the fundamental radical was a unitary core, defined by its basic structural shape, which it retained even when some of its atoms where replaced by others. For Laurent, simply listing the carbons and hydrogens in a chemical formula was not enough. What mattered was their organization. Form and arrangement, he said, were as important as composition.[60]

PARIS, COLLÈGE DE FRANCE, 1836

The focus on the internal organization of molecules had never sat well with Dumas, and Laurent's continued insistence on it soon made their relationship even worse. This did not bode well for Laurent, as Dumas was emerging as the most powerful chemist in Paris. Thénard had clearly designated him as his successor. Dumas had substituted for Thénard at the Faculty of Science in 1832, taken over the chair of chemistry at the École Polytechnique in 1835, and stepped in for Thénard at the Collège de France in 1836.[61]

In fact, the Collège de France had celebrated Dumas's appointment by having him give a series of public talks with the title "Lectures on Philosophical Chemistry." Crowds packed the lecture hall to hear the young phenomenon lay out his vision for modern chemistry, combining history and science into an ultimately moral tale about humanity's search for knowledge. Dumas's central point was that no one should let their theories outpace their observations. The great strength of chemistry, he insisted, was its "most complete faith in the testimony of the senses; its limitless confidence in experience; its blind submission to the power of facts."[62] Dumas urged particular caution when dealing with the concept of "atoms." He, like many French chemists, had long used the word "atom" as a bookkeeping device for describing the regular proportions in which things react. They liked to refer to these as "chemical atoms," and using the term did not necessarily imply that one believed in "physical atoms," or actual little particles. Indeed they often mocked those like the English chemist John Dalton, who treated atoms as real physical entities, despite the fact that they could not be seen, touched, or detected in any way. Dumas dismissed Dalton's arguments as a "vicious circle."[63] The question of atoms, Dumas pointed out, was hardly even a scientific one. It had played no role in the long experimental tradition that had done so much to advance chemistry. Its roots, rather, were in metaphysical speculation. The ancient philosopher Democritus had first proposed them as part of a system that, Dumas pointed out, bled into the morally dangerous territory of materialism and atheism.

Dumas singled out Laurent by name in his lecture, as an example of what *not* to do. He dismissed Laurent's work, warned younger chemists not to follow his example, and offered a piece of backhanded advice: "Let us carefully avoid all gratuitous suppositions. Let us always remember that there is the greatest danger in creating radical hypotheses without necessity."[64] The humiliation stung. ("My career has failed, and I will now abandon it," was Laurent's reaction when he heard.)[65] Laurent addressed the episode in the next article he wrote, revealing a rawness of emotion rarely seen in academic articles. He admitted he had been "profoundly hurt" by the public rebuke, particularly in front of such a large and illustrious audience that had included Laurent's own students. Dumas, he said, had advised him to "stick to the facts only," but he countered that he saw no point in that. "If I believed that my work could only end with finding a few new combinations, or in proving that in some body there is one or two more atoms than in another body, I would abandon it immediately." The only thing that propelled him forward was his search for the true, hidden nature of reality. The desire for an explanation, he said, "can alone engage me to follow a career in which I find so little encouragement and so many obstacles to surmount."[66]

He began organizing his ideas into a master theory, writing it up to submit as a doctoral thesis at the Faculty of Sciences. He even titled his thesis "General Considerations on the Physical Properties of Atoms and on Their Form," seemingly in direct defiance of Dumas.[67] In it, he invited the readers to visualize, in their minds, fundamental molecular forms in the shape of four-sided prisms. It was clear there would be no way to get the school's chemists, Dumas and Thénard, to sign off on it, so he sought out other members of the Faculty of Sciences, putting together a committee of two physicists, Pierre-Louis Dulong and César Despretz, and a mineralogist, François Sulpice Beudant.

Liebig offered his advice on the situation, writing from Giessen to tell Dumas he should squash Laurent like a bug. You should never ignore insects, he wrote, as they could still poison in small doses. Instead, he urged Dumas to double down on his attacks on Laurent and his ideas, which were in his eyes "true sans-culotisme." "I would never put up with such arrogance," Liebig wrote, warning that the general public, not knowing any better, might one day wind up listening to Laurent and turning their backs on Dumas. There was only one thing to do about it: "You must punch him in the stomach, such that he doesn't even think about getting up."[68]

Laurent had chosen a risky strategy, trying to make a career in chemistry while angering the most prominent members of the field. But he was about to be wealthy enough that he did not have to care. He had finally gotten the proceeds from his profit-sharing agreement: 10,000 francs each for him and Édouard, a fortune for someone who had been scraping by on eighty francs a month from tutoring. The perfume house had changed hands after the death of Édouard's father, Jean, on 15 March 1836. It kept the name Laugier Père et Fils but was now owned by a certain Monsieur Renaud, who hired Eugène Roussel, an enterprising worker three years younger than Laurent and Édouard, as the director of the laboratory. Jean's widow, Catherine Dufrayer, moved down the street to no. 8, where she continued to sell perfume and serve as a depot for Eau de Cologne from Jean-Marie Farina in Cologne.[69] Laurent and Édouard both set out on their own, hoping to multiply their newfound fortunes in the burgeoning chemical industry.

9

The Spirit of Coal Tar

PARIS, 1837

It was a cold, hard December for Laurent. His fortune—the 10,000 francs that were supposed to set him up for life—had disappeared. He had invested its entirety in a factory to make copper sulfate, a powder used to control mold, hoping to become the next Dupont or Chaptal. But the winds of private enterprise were fickle. Within months, the venture had failed, the money gone forever.[1] He had proceeded in a kind of stunned fog, finishing his thesis and scheduling its defense for the twentieth of that month. Perhaps, he hoped, the quality of the work might be enough to earn him a position somewhere. But that prospect had dimmed considerably in recent weeks, as Dumas had mounted a new offensive against him.

Dumas and Liebig had formed an alliance, coming together after eyeing each other warily as rivals for years. They closed ranks to promote their shared vision for organic chemistry, uniting to stamp out the dissenters, with Laurent first among them. In 1837, they wrote a joint manifesto entitled "Note on the Current State of Organic Chemistry," which called a new generation to arms. They admitted they had had their differences in the past, but now recognized the necessity of working together to tackle the enormity of the project ahead: the natural classification of organic matters. The most profound secrets, "the mysteries of vegetation, the mysteries of animal life," could all be explained by the same principles Lavoisier had used for the inorganic world. While promising they did not make the claim lightly, they announced that "this great and beautiful question is today resolved; there remains only to unfold all the consequences that its solution entails."[2]

They proposed a systematic plan of attack. The first step was to analyze every known organic substance and determine its chemical formula. Any analysis that they did not do themselves, they wanted sent to them, so they could verify it. The next step was to react each substance with everything they could, as Liebig and Wöhler had done with the essential oil of bitter almonds. They could then write out the reactions to see if they could find a stable radical, as Liebig and Wöhler had with benzoyl. Once all these radicals had been identified, the rules of organic chemistry would no doubt be as straightforward to follow as those of inorganic chemistry.

To tackle this massive project, Liebig and Dumas proposed the mobilization of an army of chemists, a grouping "unheard of in the history of science" that brought together all the researchers of Europe in a united effort. They would be the two heads, with Liebig operating out of Giessen and Dumas out of Paris. Each would open their laboratories to any young chemist willing to join the effort, and they would preside over them, doing their own work "surrounded by young imitators, the hopefuls of science."[3] Édouard Laugier and Auguste Laurent were not invited. Indeed, the program was a direct refutation of Laurent's general considerations. There would be no imagining what the atoms looked like in their mind, or how they came together to form prisms. No one would make crystals of the products, or try to guess at the shapes of the molecules themselves. There would be no speculation about higher levels of organization. Instead, the focus was entirely on the chemical formula.

Dumas read his manifesto on the floor of the Academy of Sciences on 23 October, as Laurent was finishing up his thesis, whose last pages noted his "indignation" at his ill treatment.[4] The oral defense, held a few weeks later, was an awkward affair. Dumas was not on the committee and did not need to be there, but he showed up anyway to barrage Laurent with hostile questions. The defense lasted over two hours, with Dumas's aggressive prosecution and Laurent's heated responses stretching into a "joust" between the two that riveted the audience.[5] In the end, there was no way to keep Laurent from passing, as he was clearly one of the best experimental chemists of his generation. But the jury still found a way to distance itself from his theoretical ideas. The tradition was to vote using colored balls. A white ball meant "very good," a red ball, "passable," and a black ball, "bad." Laurent had divided his thesis into two sections. The first, titled "Various Researches on Organic Chemistry and on the Density of Mixed Clays," was a straightforward presentation of experimental results. He got three white balls. The second, titled "On general considerations on the physical properties of atoms and on their form" was his wild ride deep into theory. It was a startlingly original vision tying together an unprecedented array of different observations. It got three red balls, barely passing.[6]

The blow hit Laurent hard, who admitted he had been "stunned" by the "painful debate" with Dumas.[7] He had known he was going against the establishment with his work, but he had been expecting to do it with the independence afforded by the fortune he had earned at Laugier Père et Fils. By the time he passed his thesis, he did not have enough money to pay for getting it printed (as was the custom at the time). Instead, he broke it into several articles for the *Annales de chimie,* where it ran under the anodyne title "Various Researches in Organic Chemistry."[8] But to get it published, he had to cut out all of the controversial parts: the speculation about the physical nature of atoms, the structures they formed, their analogy with crystals—in short, the crucial sections that lay closest to his heart.

He was ready to admit defeat. He took a job at a ceramics factory in Eich, Luxembourg—precisely the job he had hated under Brongniart. Eich was a small town whose solidly bourgeois citizens were a far cry from his Bohemian socialist friends, and he struggled to fit in. He had a birthday to celebrate not long after arriving, but he was unable to summon much enthusiasm, noting only "I am 30 years old. I have wasted my life."[9]

PARIS, RUE CUVIER, 1838

By 1838, Dumas had accumulated laboratories all over Paris: there was his original one at the École Centrale, another at the Faculté de Médecine, and one up the hill at the École Polytechnique. He had projects brewing in each one, and he was supervising an ever-expanding cadre of younger chemists.[10] Yet still he was not happy. He complained about the time he spent traversing Paris to visit them, passing through dangerous neighborhoods and paying a toll each time he crossed the Seine to check an experiment.[11] The labs themselves were far from ideal—his space at the École Polytechnique was so drafty and ill-heated that it got down to −17° C, well below freezing, inside the building. Even if he had been willing to bear it, none of his operations would have worked at those temperatures.[12] Jules Pelouze, who worked across from Dumas at the École Polytechnique, had spent a summer with Liebig in 1836 and returned to report that he had a new lab of incomparable caliber. "Giessen, Giessen, ah!" he would exclaim in the following months, transported by the memory of it.[13]

Liebig's new space had been hard won. The University of Giessen had been happy to leave him in the abandoned barracks, where he outfitted the lab from his own pocket and complained to deaf ears about the poor ventilation and inadequate heating. It was all so much that he had collapsed from exhaustion, suicidal and unable to work. Wöhler suggested he had "*Hysteria chemicorum*," a "specific illness of the chemist" brought on by mental overwork and exposure to strong smells.[14] Liebig spent months recuperating in a spa, and from there he wrote a letter to the university chancellor telling him he would pack his equipment and leave if forced to return to those conditions. The University acquiesced, beginning a series of major renovations in 1835.[15] He began accepting twenty new students at a time to join him, forming a tight-knit group with a fine-tuned hierarchy and division of labor. "Turning out the most audacious results in factory-like fashion," was Liebig's description.[16] Dumas sensed himself being outstripped. He and Liebig had vowed to build an army of young chemists together, but the troops were turning out to be Liebig's. "How very fortunate you are," Dumas wrote to Liebig, "to have a battalion of eager chemists at your disposal. I hope someday to offer you something comparable, but for the moment I am far from that."[17]

Dumas dreamed of a Giessen-like lab of his own, and he happened to have the wealthy father-in-law to bankroll it. He and Herminie had been living in her family's grand apartments at the Museum of Natural History since their marriage, but in the spring of 1838, Alexandre Brongniart began building a new home for them across the street that would double, on its bottom floor, as the finest chemistry lab of Paris.[18] Dumas stocked it with equipment from his lab at the École Polytechnique, and by that fall began welcoming young chemists to work at his new, private lab. In November, the city renamed the street it was on after the recently departed Georges Cuvier, and before long the mention of "rue Cuvier" became synonymous with Paris's most sought-after lab space.

Dumas soon had some thirty students working under him, drawn not only from France but all over Europe. Together they scoured nature for its most remarkable plants, to wrest from them their chemical formulas. Dumas himself continued his work with lavender and lemon, and started a new analysis of meadowsweet flowers, often used to flavor mead and thought to have medicinal powers. He directed the younger chemists around him in an orchestra of smells and smoke. In one corner, his student Eugène Peligot took up the study of cinnamon. In another, Z. Delalande analyzed coumarin, the fragrant, vanilla-like substance extracted from the tonka bean, while Francis Scribe worked on St. Benedict's thistle, used in tinctures to treat colds.[19] His student Malaguti studied bezoars, the "stones" of organic material traditionally used as an antidote to poison.[20] Chemists came from Denmark and Poland to study valerian root, peppermint, willow bark, or the use of the Japanese wax plant for candles.[21] Even Jules Pelouze, just appointed assayer to the mint, used Dumas's lab to work with him on the analysis of black mustard.

Busiest of all was a pair named Auguste Cahours and Charles Gerhardt, who were working together on a broad survey of essential oils. Cahours had replaced Laurent at the École Centrale and was Dumas's clear favorite and heir apparent. Gerhardt was an Alsatian from Strasbourg who had fled his father's business to study chemistry in Giessen. After eighteen months there, he went to Paris on Liebig's recommendation and found a spot in Dumas's lab.[22] With Cahours, he analyzed the essences of chamomile, fennel, star-anise, valerian, cumin, elfwort (one of the ingredients of absinthe), and more.[23]

Dumas had hoped that once all these substances were analyzed and assigned precise chemical formulas, the relations between them would be clear and the underlying laws of organic chemistry would be revealed. But nothing of the sort happened. If anything, the situation became murkier and more confusing. The only thing that was clear was that molecules could not be treated as having a positive side and a negative side, attracted to one another by electrical forces. So Dumas changed course, dropping the idea that radicals functioned as one component of a dualistic pair, and instead focusing on them as a unitary fundamental, which he called a "type." These "types" closely resembled Laurent's fundamental nuclei, but Dumas ignored that implication, and instead insisted that he had proposed an entirely new "type theory."[24]

Liebig was having no better luck up in Giessen, and was, if anything, even more exasperated by the muddle of results his battalion of students provided him. "The loveliest theories are being overthrown by these damned experiments; it is no fun being a chemist any more," he complained.[25] And so he left the field entirely, publishing his last paper on organic chemistry in 1839.[26] He turned instead to "a completely different side of science"—the practical applications of chemistry for improving agriculture.[27] He had discovered how to artificially fertilize crops after noticing the role of nitrogen in plant growth, and was quickly becoming richer and more famous than ever.

Liebig was not impressed with Dumas's type theory, and he let the world know with his characteristic cruel humor. Wöhler had sent him a satirical piece mocking Dumas in 1840, intending to circulate it among his friends. But Liebig published it in his journal, *Annalen*, which was by now the primary German-language vehicle for chemical news. He gave it by byline "S.C.H. Windler," or "swindler" in German, but there was no mistaking Dumas as the target—its title referred to his theory of types, and it began with the hearty "Monsieur!" Dumas was known for in his lectures.[28] It went on to recount how he had taken a piece of cloth and replaced all of its atoms one by one with chlorine, until it consisted of nothing but chlorine atoms, all the while, according to Dumas's theory, retaining its original type. The perfectly bleached cloth, it was noted, was now on sale in London and "very much sought after and preferred above all others for night caps, underwear, etc."[29] Dumas was unamused. But he ignored Liebig's provocation for now. He had other issues to deal with. Although his punch to Laurent's stomach had landed hard, Laurent was refusing to stay down.

BORDEAUX, 1839

Laurent's luck in business affairs had hardly improved after arriving in Luxembourg. The ceramics factory went under, leaving him without even the job he had so begrudgingly accepted. But his luck in love had come through. He was married now, to Anne-Françoise Schrobilgen, known to everyone as Francine. The Schrobilgens had been one of few families in Eich to appreciate the Parisian's strange charms, although their first meeting had been a near disaster. Laurent had come to dinner as the guest of the director of the ceramics factory, Norbert Metz. After the meal, everyone tried to convince one of the host's twin daughters, Francine, to sing for them, but she resisted. Laurent, for encouragement, promised that if she sang a song, he would, too. When his turn came, he sang a satiric ditty "J'ai du bon tabac," which bragged about the fine contents of his snuff box before concluding that it was "not for your ugly nose."[30] No one appreciated the joke, and Laurent soon made his excuses to leave, convinced he would never return. The Schrobilgens, however, lived outside the city's main gate, which, as Laurent soon discovered, was locked every night at 10 o'clock. Unable to get home, he made his way back to his host's, where there was nothing to do but stay up all night drinking. The father, Mathieu, worked as a clerk in the city's courts, but it turned out that his real passions were composing poetry, playing the violin, and writing subversive journal articles that he signed "schro." The family became his closest friends in town. By July of 1838, he and Francine were married.

The two soon left for Bordeaux, where he had accepted a job teaching chemistry. He had done it out of desperation, in need of the money, but was pained to be moving even farther from Paris, a full five days of travel by coach.[31] In France, scientists saw provincial appointments as nothing less than professional death. Not only was he stuck far from where things were happening, but the university structure of France was set up so that research was really supported only in Paris. According to official instructions, the laboratories of provincial universities were designated strictly for class preparation. Private research by professors was, if not out and out prohibited, at least frowned upon.[32] Bordeaux, in fact, did not have a laboratory at all when he arrived. The university had just added the Faculty of Sciences, and Laurent held their first-ever chair of chemistry. Housed with the rest of the sciences in a makeshift annex to the Hotel de Ville, Laurent set up his laboratory himself, equipping it out of his own threadbare pocket. The tiny, dingy space stood in unflattering contrast to Dumas's brand new lab, which brimmed with a team of trained chemists, eager students, and numerous assistants to help him. Laurent had no one, and the prospect seemed impossible. "But how am I supposed to manage it alone?" he lamented. Without even the most basic form of assistance, he lost an hour every morning simply sweeping and cleaning glassware. "Ah scoundrel! Dog of a job! Cursed Provinces! What! I have nobody here to help me!"[33]

Laurent had to buy all of his reagents himself, and every purchase, he admitted, had to be weighed against what would go missing from his dinner's stew pot that evening.[34] He returned to one of the cheapest substances he had worked with: coal tar. Bordeaux had just established gas lighting in its streets shortly before Laurent arrived, which meant its municipal gasworks had cisterns of coal tar refuse there for the taking. He had never forgotten the intriguing parallels between bitter almonds and naphthalene, the coal tar product whose smelly production made it ill-suited for the back of a perfume shop. Now alone in his lab with no one to bother, he plunged back in.

At the center of his work was a substance he called "the spirit of coal tar." He gave it that name, with its alchemical echoes, as a deliberate form of analogy. Just as the distillation of wine produced the spirit of wine, so too did the distillation of coal tar produce its own kind of spirit. This was not an entirely novel move. When Laurent had worked with Dumas at the École Centrale, they had worked on the distillation of wood (this was a "destructive distillation," done at high heats inside a dry flask), which produced a substance known as the "spirit of wood." Dumas analyzed the substance, and determined it was a hydrate of what he called *methylène* (which he got from the Greek words for "wine" and "wood"). In naming it, he followed the example of Lavoisier, who had identified the spirit of wine as a hydrate of sugar, and named it alcohol. Dumas thus called the "spirit of wood" methyl alcohol, shortened to methanol. In this same vein, Laurent asked what it was the spirit of coal tar could be a hydrate of. The answer? *Phène*—the substance he had derived from the essence of bitter almonds, that Mitscherlich had called benzin and Faraday the bicarburet of hydrogen. To continue the analogy with alcohol and methanol, the spirit of tar became "phenol."[35]

The spirit of tar was his mystery to be solved. What was it? He noted that it tasted and smelled like creosote, another coal by-product that railroads sometimes used to preserve wooden ties and pharmacists prescribed for skin disorders, wounds, and toothaches. Wanting to compare, he sought out any friends who had toothaches, pressed them to treat it with phenol, and reported back that it seemed to have the same numbing and cleansing effect as creosote, "strongly attacking the skin of the lips and the gums."[36] Laurent also wondered if it could be that same substance the Hamburg chemist Friedlieb Runge had found in coal tar in 1834 and named "carbolic acid." Laurent suspected the "spirit of coal tar" was the pure form, while both creosote and carbolic acid were mixed with contaminants. But the case was not yet clear.[37]

For Laurent, it was all part of his bigger theoretical scheme, which he held onto in spite of the rough reception it had received. He remained convinced that the key to understanding organic molecules was their organization, analogous to crystalline form, and that this applied to coal as well as living plants. Naphthalene and the essence of bitter almonds formed two parallel series, running side by side and separated only by the two additional oxygen in the fundamental nucleus of bitter almonds. Would it be possible to jump from one series to another, to transform one into the other? It would involve, he saw, removing two oxygens from the essence of bitter almonds, or "deoxydating an organic acid," something no had ever done before.[38] But he persisted, eventually creating a new substance from bitter almonds that he named stilbene. When he examined its crystals, he found they formed rhomboidal prisms resembling naphthalene—proof, in his eyes, that he had changed the organization of the fundamental nucleus. His manipulations, he said, "confirmed the rule that *I alone* have established."[39] He had bridged the divide between coal and living things, transmuting one into the other. Of course, he started with something very expensive (the essential oil of bitter almonds) and ended in the direction of cheap coal tar, leaving much room for improvement in the eyes of any alchemist.

Laurent began working with indigo, nicknamed "blue gold" for the astronomical price it could fetch.[40] Demand had soared recently, when it became the dye of choice for a popular new cloth out of the south of France called "serge de Nîmes," shortened into denim in English. The dye was expensive—its only source was the tiny leaves of a flowering shrub that grew only in tropical climates. Laurent got his hands on some and began running it through the same operations he had performed on the bitter almonds, treating indigo as a hydrocarbon nucleus, the fundamental radical of its own series. He announced that he had made several new molecules, such as one he named isatine, which could also be found in woad, another plant used for blue dye.

His successes roused the attention of Dumas, who had previously worked on indigo but had never gotten very far. Unwilling to cede any territory to Laurent, he took up the subject again, writing a long, comprehensive overview for the *Annales de chimie*.[41] He essentially copied

Laurent's approach, calling indigo a "type" or "one of these hypothetical radicals" that anchored a series of its various derivatives.[42] But he gave Laurent no credit, and continued to disparage his larger theory about the physical organization of molecules. Laurent was livid. Dumas had named his essay "Fourth memoir on types," and Laurent took the title as an inspiration for his own series of snarky articles, published in a rival journal, culminating in "Thirty-first memoir on types or derived radicals (which were not invented by M. Dumas)."[43]

Laurent had never been more isolated. But he worked in a fever of activity, producing paper after paper, and by 1841 had notched a remarkable victory. Starting with nothing but some "spirit of coal tar" he had produced indigo bitters (or picric acid, as Dumas had renamed it), a bright yellow powder derived from indigo.[44] He had crossed the boundary between coal and living plant, hopped from one series to another, this time in a direction the alchemists would have approved: from base substance to blue gold, or at least one of its derivatives.

GIESSEN, 1843

The rest of Europe did not quite know what to make of Laurent. There was no denying the importance of his experimental results. Berzelius, now something of an elder statesman in the field, wrote a report every year summarizing the current state of chemistry. In 1842, he gave top billing to Laurent's work on indigo, calling it the most important contribution to organic chemistry in many years.[45] But he could not bring himself to follow Laurent's theoretical speculations, which he called "a chaos of oddities."[46] Liebig was similarly impressed with his experimental output. He had translated twenty-six of Laurent's papers into German to publish in his journal, the *Annalen der chemie und pharmacie*, the most of any French chemist after Dumas, who had twenty-nine.[47] But he also kept his distance from Laurent's bolder claims. "He's crazy," confided the young chemist Charles Gerhardt to Liebig in 1842.[48] And Liebig did not disagree, advising Gerhardt that he must by all means avoid making theories, or else "you will destroy your future and irritate everyone, like Laurent."[49]

It stung, for Laurent, to see his ambitious vision of a new chemical system dismissed as mere eccentricity. He complained that Liebig could be condescending, treating him "like a little boy."[50] He was particularly upset when Liebig had attacked his work on the spirit of coal, writing up a brochure accusing Laurent of stealing his ideas from Friedlieb Runge without credit. It was dispiriting, not so much for the accusation, which was easy to disprove given that Laurent had cited Runge twenty times, but for the clear inference that Liebig had not bothered to read the paper with any care.

He determined to confront Liebig in person as soon as the school year ended. He and Francine would already be in Luxembourg, and Giessen was only 150 miles farther, hardly anything after the long journey across France from Bordeaux. They made the trip to visit Francine's family in Eich every summer, and their son, Hermann, had been born there in 1841.[51] The boy was now a toddler, able to walk, and Laurent left him with his wife and in-laws to go set the record straight with Liebig.

The confrontation was anticlimactic. Liebig dispersed insults so freely over his career that he scarcely remembered the particular instance. He admitted he was wrong and joked that whenever he wrote another attack like that, he would lock it in his desk for six months first to let his head cool down.[52] He then invited Laurent to stay and give a public lecture on his work. While Liebig himself had given up organic chemistry, several of his students still worked on the topic, and after reading some of Laurent's recent work, they were eager to hear him speak.[53]

Laurent's experiences in Giessen gave him a glimpse of the opportunities for camaraderie and efficiency he had missed by working alone. Liebig's lab had never been so busy, with sixty-eight students auditing his chemistry and pharmacy classes, and a battalion of assistants working under his command.[54] The newly renovated lab space was palatial, with something Laurent had never seen before: ventilating hoods that would draw away the fumes and smoke that accompanied experiments. Loathe to leave, Laurent began collaborating with one of the youngest chemists in the lab, August Hofmann, who shared his romantic temperament and Bohemian style, sporting a similarly rakish beard. His father, an architect,

FIGURE 26. Justus Liebig's laboratory in Giessen in 1842. August Hofmann is on the far left, holding a kaliapparat.

had been in charge of expanding Liebig's laboratory in the 1830s. He had at one point confided his worries to Liebig about his son. The boy, he revealed, loved only literature and art, and intended to study, of all things, philology at the university. "Give him to me," Liebig reportedly responded, "and I'll see what can be made out of him."[55] Liebig set him on the path to chemistry, and Hofmann received his doctorate in 1841. When his father died in 1843, Hofmann moved in with Liebig and became his personal assistant.

Like Laurent, Hofmann's first project had been on coal tar, which Liebig had assigned to him after the owner of a nearby tar distillery had sent in a sample to be analyzed.[56] One of the products he produced, by treating the coal tar with lime, was a pale yellow substance that reeked of rotting fish. He was not the first to discover it—Runge had already found it and named it kyanol—but after working out the chemical formula,

Hofmann began to suspect that it was identical to two other previously identified substances: crystallin, produced by the dry destruction of indigo, and aniline, produced by treating indigo with caustic potash.[57] If these really were all the same thing, it posed another tantalizing connection between coal tar and living plants. Hofmann shifted his work to indigo, reading Laurent's latest papers with interest.

Suspecting that the "spirit of tar" was presiding over the affair, Laurent and Hofmann set out to make their own aniline, starting with Laurent's "phène" (or benzene). They whipped up their concoction together, then placed it in a sealed tube to heat for three weeks, patiently waiting for the transformation to occur.[58] Laurent saw the operation as a test of his entire theoretical system, noting "It is in starting from views that I had on the constitution of organic bodies that we had discovered this procedure, and not by luck."[59] At the moment of the big reveal, when they had removed the tube and were about to break it open, Laurent called Liebig over, to witness the test of his prediction. Sure enough, the tube, when cracked open, was full of aniline.

It was as close to alchemy as one could hope to get. The result itself was unappealing—it looked and smelled like what someone might get if they squeezed out the liquefied innards of rotten fish. But the implications were wondrous. The quest for indigo drove a pitiless global enterprise of overseas plantations relying on slave labor in South Carolina and the Caribbean, and brutal coercion in India. Laurent and Hofmann, on the other hand, had gotten one of its derivatives from a waste product, something so cheap and plentiful you could buy a year's supply for mere pennies.[60]

What was even more remarkable was that aniline had all the characteristics of an organic base, or alkaloid, those compounds like nicotine and caffeine that acted so wondrously on the body. Laurent and Hofmann realized this right away. They concluded their paper with the claim, "We foresee the possibility of making quinine, morphine, etc. if we can discover the corresponding compounds."[61] And what a possibility that was. Opium, tobacco, coffee, tea, cinchona—these were the products that built empires and shifted the populations of the globe. England and China had barely ended their war over opium. To reproduce their active principles cheaply in a lab would change the world.

With the summer vacation over, Laurent and Hofmann parted ways. Hofmann remained in Giessen for the moment, nursing his own growing ambitions. Laurent returned to Bordeaux and the solitude of his empty lab. Francine had stayed in Eich with Hermann, who could play in his grandparents' sprawling gardens.[62] Laurent spent all of his time in his lab, surrounded by the new objects of his research: opium, cinchona, and tobacco had joined the coal tar and bitter almonds on his shelves. He had returned from his work with Hofmann in Giessen fired by the possibility of creating artificial alkaloids from coal tar. After buying some opium, he had isolated morphine from it, and corrected Liebig's chemical formula for the substance, which had been slightly off.[63] But what interested him most lay beyond the chemical formula—in these substances' ability to act upon the human body, whether the pain-relieving bliss of morphine, the fever-breaking tonic of quinine, or, Laurent's personal favorite, the stimulating buzz of nicotine.

Above it all hung the dream of transformation, in which the active principles of plants and the world of petrochemicals would transmute and combine. "I no longer think I dream,"[64] he said, reporting that for the past eight days and eight nights, he has seen before his eyes only phantoms of his cinchonine products metamorphizing from one substance to another. He took up his old habit of painting, creating phantasmagorical scenes of impossible creatures. In one, a demon calmly turns the fire spit while a cloven assistant washes at a trough and a bespectacled monkey leans studiously over a book in the middle of the room. On the wall, the pots and urns have come alive, with grimaces ranging from jolly to dyspeptic. A bird-legged woman and a violin with a dunce cap fly about the ceiling, on a broomstick and umbrella, respectively.

The contents of his flasks were stranger still. Parisian chemists remarked upon the "bizarre compounds" his theories had led him to, as his transformations created one strange substance after another.[65] He had perfected a technique for making aniline by treating an oxide of the spirit of

FIGURE 27. Auguste Laurent painted this watercolor in 1844. It presents an interior, perhaps a laboratory, with fantastical creatures.

tar with ammoniac acid. By the summer of 1844, he decided to try the same process on an old, familiar substance: bitter almonds. "I don't know what madness pushes me in spite of myself towards the essence of bitter almonds," he wrote.[66] But whatever it was, it worked, producing a new, artificial compound with the properties of an alkaloid. He named it amarine, and soon produced a second, one, lophine, to go with it.[67] Would these artificial alkaloids have wondrous properties of their own?

PARIS, COLLÈGE DE FRANCE, 1844

The chemists of Paris paid little attention to Laurent and his strange creations. But upstairs at the Collège de France, the physicist Jean-Baptiste

Biot took note. After his disagreement with Dumas, Biot had continued his work on organic fluids, shining beams of polarized light through them to see if they rotated. And chemists continued to ignore him. "We know that Biot has strongly recommended to chemists to direct their attention to the optical activity of liquids," the journalist Quesneville reported, although, "Up until the present, few people have made use of this reaction."[68] Nonetheless, Biot was more convinced than ever that artificial creations could not replicate nature.

The recent wintergreen affair had only confirmed it in his eyes, despite the triumphal claims of Auguste Cahours that he had been able to synthesize it artificially. Cahours had begun working on wintergreen years ago, when he and Charles Gerhardt teamed up in Dumas's lab on rue Cuvier. Wintergreen was a novelty then. It came from the flowers of the Gaultheria shrub, native to New Jersey, and had started becoming popular in perfumery as trade with America had increased. He and Gerhardt had together perfected techniques for separating out the often complex mixtures responsible for fragrance in plants. They had begun quarreling, however, and now had a tense relationship. Cahours had remained close with Dumas, now serving as his assistant at the École Centrale, while Gerhardt had left Paris in a kind of exile, accepting a teaching position at Montpellier.

In 1844, Cahours took up the project of synthesizing artificial chemicals. He started with the spirit of wood, the clear liquid that Dumas had named methanol. By combining it with salicylic acid, he produced a new compound, methyl salicylate, that smelled like the natural oil of wintergreen and had the same chemical formula. Dumas backed him up when he reported Cahours's result to the Academy of Sciences, claiming that Cahours's wintergreen was "easy to produce artificially" and "possesses exactly all the properties of the natural oil."[69] Dumas, however, was wrong. There was one property they did not share. The natural oil from the Gaultheria plant was optically active, and the artificial oil was not. Discerning noses also swore there was a difference in how they smelled.

But what, exactly, was the nose capturing that all other chemical tests could not? The strange case of caraway and spearmint deepened the mystery. Caraway seeds have an utterly distinctive scent, deep and spicy, giving rye bread its particular flavor. Spearmint has an entirely different odor palette: fresh, sweet, and light. Yet as chemists were able to isolate the parts of these plants, they found the two substances shared what seemed to be the exact same factor responsible for their odor. Chemically identical in every way, they could be distinguished only by smell and a curious fact of polarization: oil drawn from caraway rotated polarized light to the right, and oil drawn from spearmint rotated light to the left. Was there information carried in smell that could not be captured by the chemical formula? It bore a resemblance to the *esprit recteur,* an invisible, ineffable factor that governed the human experience of scent.

Biot was convinced that the answer lay in the physical structure of the molecule, which somehow affected how these substances acted upon the human body. Smell was often the first, most immediate form of action upon the body, and one that had presided over Biot's thirty years of investigation into essential oils. He had recently begun investigating another class of molecules that acted even more dramatically upon the human body: alkaloids. They were generally known as the "active principles" of a substance—the part of opium responsible for its stupefying effects, the part of cinchonine that reduced fevers, and so on—and Biot wanted to see if his polarimeter detected any optical activity.

Biot collaborated with Apollinaire Bouchardat, the chief pharmacist at Paris's top hospital, the Hôtel-Dieu. Bouchardat had installed a polarimeter in the hospital in 1840, which he used to test for the presence of sugar in his patients' urine. It was the first laboratory diagnosis for the condition, and it earned him the title of founder of the field of diabetology. But Biot was more interested in his access to the hospital's pharmacy stores, which ranged from medicines to intoxicants to deadly poisons. In 1843, following the procedures of Biot, Bouchardat began to put one "active substance" after another into his polarimeter to see what happened. And they proved very active, indeed, showing "very marked and very diverse" rotations.[70] Bouchardat was particularly impressed with quinine, which "possesses such beautiful optical properties" that he envisioned using polarimeters to test the purity of samples for the colonial army, thereby reducing the "pernicious and murderous fevers" faced by "our soldiers in Africa."[71] He communicated his findings to Biot, who presented them to the Academy of Sciences.[72]

When Biot read, soon after, that Laurent had succeeded in making artificial alkaloids, he saw an opportunity. The alkaloids from plants were all optically active, but what about the artificial ones? Biot offered the use of his instruments to Laurent to test them, and Laurent set out for Paris as soon as his school year at Bordeaux ended. They soon had their results: neither were optically active.[73] This led them to another series of investigations, in which they started with pure samples of cinchonine and strychnine and then modified them, with Laurent swapping out hydrogens for chlorine to progressively change them into something different. At each step along the way, he tested the optical activity of his new creation, as well as the potency of its active principle. One of Laurent's old students from his private tutoring days, Jules Maisonneuve, was now an assistant at the Hôtel-Dieu, and he ran tests with the strychnine derivatives, feeding them to dogs and timing how quickly they died. Their conclusion: the "action" of the molecules on the animal economy diminished hand in hand with its "action" on polarized light, as the substance became gradually less natural.[74] Laurent did one more comparison in Biot's polarimeter, this time between nicotine and the aniline he had prepared with Hofmann. His discovered that nicotine extracted from plants was optically active, but aniline, created synthetically, was not, puncturing his dream of easily producing alkaloids from coal tar. He ended his paper with a sobering reflection. "If natural substances, or some of their products of transformation, alone possess rotary power, we must despair of reproducing nicotine by artificial means."[75]

10

The Study of Things That Do Not Exist

LONDON, 1845

Laurent was running out of friends. He tried writing to August Hofmann, the young chemist who had been so happy to see him in Giessen, but by the fall of 1845, Hofmann was no longer returning his letters. "May they now go hang themselves!" wrote Laurent, of Hofmann and every other chemist who ignored him.[1] Hofmann had enjoyed a meteoric rise since the two had worked together in Giessen. He had proven himself a man of vast ambition, both for his personal career and for the power of chemistry. His first idea was to try to marry Liebig's daughter, but she would not go along with the plan. He settled instead for the niece, although the move did not produce the intended consequences. He had procured the engagement while Liebig was out of town and arranged to teach in Bonn to earn enough money for married life. When Liebig returned he pronounced himself "painfully moved" that Hofmann should have done so "without consulting me and without telling me about it."[2]

But fortune soon shone anew, as he found an even faster path up the ladder of ambition. The English had been watching the success of the laboratories of Dumas and Liebig, well aware that England had no equivalent institution for training chemistry students. Queen Victoria's consort, Prince Albert, pushed for the creation of a Royal College of Chemistry in London, securing funds from over 900 donors. But the college still needed a director, and no one in England seemed able to fill the role. The question was still unresolved by the summer of 1845, when the queen and prince consort paid a state visit to the Prussian court. Prince Albert, German by birth, had attended the University of Bonn before he was married and asked, for the sake of nostalgia, to revisit the hall where he had lived as a student. This happened to be the building where Hofmann was setting up his new laboratory for the classes that would start in the fall. He ran into the royal entourage and gave them an impromptu demonstration of several dramatic chemical reactions. Albert hired him on the spot as director of the Royal College of Chemistry (the minister of education was also in attendance to clear up any paperwork).[3] Hofmann began repacking the bags he had just unpacked. "Once in a lifetime each of us is offered a very special chance, and the daring man seizes it," he wrote to Liebig, as he set sail.[4]

The promises from England were staggering: an income ten times what it would have been in Bonn and enough laboratory space for forty assistants, students, and technicians.[5] Within a month, Hofmann was at work in London, with a first cohort of twenty-six students working under him. He had moved to the land where coal was king, and so he continued to devote much of his research program to exploring its chemistry. By that fall, he had discovered a new method of producing benzene from coal tar. From there, he tried to make more aniline, but he was struck by one of the intermediate products produced along the way, one that he called nitrobenzene. In his laboratory full of things that smelled like rotten fish or gasoline, it actually smelled quite nice, like almonds. In fact, it was indistinguishable from the essential oil of bitter almonds, the crucial ingredient of so many soaps and perfumes. Hofmann sensed an opportunity.

He assigned the study of the substance to one of his students, Charles Blachford Mansfield, a rector's son who had just joined the Royal College. A shy, teetotal vegetarian, Mansfield had struggled with depression for years before taking up the study of chemistry.[6] While he still struggled to control his dark moods, he threw himself into this task without reservations. After some tinkering, he found a way to make massive amounts of nitrobenzene quickly and cheaply by treating benzene with nitric acid. He took out a patent on the process, filed on 11 November 1847, and began preparations to set up his own business as a manufacturer. Hofmann looked on with approval, highlighting its application to perfume in his annual report on the achievements of the Royal College. As he put it, "this perfume may now be procured from coal-tar in tons, if required, with the greatest facility and at a trifling cost."[7]

But Mansfield's patent did not extend to France, and that soon proved disastrous for him. A detailed account of his process had appeared in the French journal *Le Technologiste* in October 1848, describing the transformation of coal tar into an essence with a "smooth and aromatic scent."[8] A couple of weeks later, on 16 November, a science-savvy pharmacist named Claude Collas made his way to the Academy of Sciences in Paris and deposited a "paquet cachet," or sealed packet of his secrets.[9] This was a common practice at the Academy, used by people who wanted to establish priority for an idea but not yet make it public. The Academy would mark the date it received the packet and publicly reveal its contents only if the bearer requested it.

Collas was a master of the marketing gimmick. He had originally become rich back in 1840, when he was just turning thirty, by pioneering a method of selling branded medical pastilles in his pharmacy on rue Dauphine in Paris. He mixed his medications into sugar pastes and pressed them into molds, leaving the pills embossed with his insignia. He was the first in France to emboss pills on an industrial scale.[10] He had then teamed up with the chemist Jules Pelouze to build a large factory on the outskirts of town for making benzene from the used-up coal tar from the Paris gasworks.[11] He sold it as a cleaning liquid to remove grease and resin from all kinds of cloth, pitching it as a replacement for turpentine and lemon oil, both botanical essences that were significantly more expensive. His talent for self-promotion again proved crucial, as he called it "Benzine Collas" and took out huge advertisements that stretched across the pages of Paris's most widely read newspapers. He was so successful that, in

France at least, people used the term "Benzine Collas" to refer generally to benzene itself.

The big secret he deposited at the Academy of Sciences was little more than the process Mansfield had already described in the article in *Le Technologiste.* But Collas invented an alternative narrative that he tried to pass off. In it, he dated his discovery to several months earlier, in March 1848. He had been out at his factory, he said, overseeing the production of benzene, when he decided on a whim to add some nitric and sulfuric acid to the benzene, claiming he'd been inspired by reading about some Swiss chemists who had treated cotton with these chemicals to produce an explosive called guncotton. His own efforts yielded a solid yellow button at the bottom of the vat. (We may note that nitrobenzene only turns solid under 5.7° C, which is perhaps why he had to move his fabricated date all the way to March, to make sure it was cold enough). He was, he reported, wondrously surprised when he first smelled it, then tasted it, and found it pleasant and sweet.

He named his new substance Essence of Mirbane, and pronounced it a perfect replacement for the essential oil of bitter almonds—and eight times cheaper to produce. Not everyone was fooled by the origin story he presented. Fellow chemists pointed out that it seemed to postdate Mansfield's work, and that he had consulted with Dumas's student Jules Pelouze to get the details right, a fact that Collas himself tried to obscure.[12] Mansfield complained that Collas had "taken advantage of the chaotic condition in which our patent laws were a few years back, and has turned another man's, to wit my, labour to good account."[13] But these reservations had little effect on the commercial success Collas soon enjoyed. Whether or not he discovered it on his own, he was certainly the first to make, market, and sell an artificial perfume. His new product got the full marketing workup. Ads soon appeared for the soaps and fragrances now scented with mirbane, claiming it was indistinguishable from the real thing, enjoying all of the aromatic properties of bitter almonds. Indeed, Collas claimed, it was likely even better for you because real bitter almonds contained ample amounts of cyanide alongside their essential oils. Mirbane was free of that danger, and Collas even envisioned using it to flavor drinks and liqueurs.[14] What could possibly go wrong?

MONTPELLIER, SPRING 1845

Charles Gerhardt had been warned in two different languages to stay away from Laurent since he first began his study of chemistry. Born in Alsace, he had grown up speaking both French and German at home, and his efforts to flee the family business brought him first to Liebig's lab in Giessen, where he spent eighteen months in 1836 and 1837, and then to Dumas's lab in Paris, where he arrived in 1838 soon after Laurent had left for Bordeaux. The rumors he heard painted Laurent as fractious and delusional. And Gerhardt had avoided him, even attacking his work to try to gain favor with Liebig and Dumas.

But by 1845, he was having doubts. He had taken a position similar to Laurent's, at a provincial university in Montpellier. Having set out for the post with high hopes, he saw his dreams evaporate within the space of six months as the extent of his scientific exile revealed itself. He had no money or support for research, and when he tried to do some anyway, the other professors reacted with jealousy and suspicion.[15] Over the school break, he found both himself and Laurent in Paris and asked a mutual friend, Gustav-Augustin Quesneville, who edited the *Revue scientifique*, for an introduction. Gerhardt tested him a bit at first, opening with a criticism of some of Laurent's work.[16] But Laurent was even-tempered in his response, and they soon worked out what the issue was.[17] He was hardly the difficult madman Gerhardt had been led to expect, and the two soon began to bond over the difficulties of doing research in the provinces. "I am dying of boredom here," Laurent wrote to his new friend.[18] They commiserated over the paltry lack of funds, as Laurent said that every day he had to calculate his finances to know "whether to put one or two sugar pieces in my coffee."[19] Gerhardt complained of his single, barely qualified assistant, and Laurent replied that he was lucky to have that, and that he himself was completely alone "in the middle of my retorts and evaporating dishes, tending to first one and then the other."[20]

Slowly, Laurent gained a convert to his cause. He persuaded Gerhardt to think about atoms as real, physical things, and the two pioneered the language of atoms and molecules as we know it. Gerhardt soon adopted the idea of a fundamental nucleus defined by its shape and the series that they anchored. They began to work together on opium derivatives, bitter almonds, and the spirit of coal tar; indeed, it was Gerhardt who popularized the use of "phenol" to refer to the spirit of tar.[21] Now that he had at least one other person to talk to, Laurent plunged even deeper into his own world. He invented an entirely new nomenclature, with his own idiosyncratic rules, intended to "indicate the number and the arrangement of the atoms."[22] He admitted that "At first, some of these names might appear rather bizarre . . . one may laugh at words such as chlophénèse, bronaphtise, indinase, chlorinidinèse, arethasum, etc."[23] But he insisted that they were a better reflection of the underlying reality.

Together they made plans to take on the chemical establishment and overturn its orthodoxies. Gerhardt was nine years younger than Laurent, and the bolder of the two. Laurent preferred to wage what he called "guerrilla warfare," taking strategic shots at his adversaries but also knowing when to hold fire. Gerhardt's battle plan, on the other hand, was to charge across the open field with banners unfurled, broadly denouncing his enemies' ideas. It was now Liebig's turn to warn Laurent to stay away from Gerhardt, reminding him in a letter that he had much more to lose in the association, as Laurent at least had the reputation of being a first-rate experimenter.[24] But the two pressed on in what Liebig would call their "monstrous alliance." They looked around for additional allies, wondering if perhaps Hofmann could "enter into our league," but he declined, saying he did not see the point.[25] They reached out to Édouard's old partner, Kramer, now in Milan.[26] Édouard himself could be of little help. He had lost his fortune only slightly more slowly than Laurent, taking a few years to go bankrupt before returning as the laboratory director at the perfume shop. No longer working for his parents, he had less latitude to spend his time on chemical research. Laurent and Gerhardt worked on as an isolated pair. One problem was where they would publish. Of the two main journals, Dumas controlled the *Annales* and Liebig the *Annalen*. Although they published articles there, "We cannot say everything that we want," Laurent confided, as they had to stick to the experimental results and leave out their broader vision.[27] They made do by publishing in Quesneville's journal but were soon discussing founding their own journal as the mouthpiece for their own school of chemistry.

First they had to get out of the provinces. Gerhardt proposed that they rent a little room together in the Latin Quarter of Paris.[28] Dumas controlled the city's labs as tightly as before, having by now collected all of Thénard's old positions. Laurent was hoping to "try a rapprochement" with him.[29] After all, Dumas had showed some signs of détente by nominating Laurent for the Grand Cross, the first rank of the French Legion of Honor, and also making him a corresponding member of the Academy of Sciences that April. Usually reserved for scientists who lived outside of Paris, the award meant that while Laurent could not attend meetings, he could submit memoirs through one of the full members. Perhaps, Laurent hoped, there was enough goodwill to help him secure a position in Paris. He impatiently awaited the end of the spring semester so that he could set out. Gerhardt, who lived farther south, finished before Laurent, so he intended to stop in Bordeaux, and they would leave together for Paris. "Come see my little kitchen and my products," wrote Laurent at the end of June. "We will travel together to the imperial city, and there, under the influence of tobacco smoke and a purer air than that of our laboratories, we will make projects in the air."[30]

PARIS, SUMMER 1845

Their time together in Paris was far too short. They arrived on 20 July, and by August, Gerhardt had to return to Montpellier to prepare for the fall. Laurent decided to stay, arranging to take a leave of absence from his teaching position at Bordeaux. He would only get half his salary, however, and it did not go far in Paris.[31] He searched for an apartment where he could bring his wife and child when they joined him in the fall after their usual summer in Luxembourg. He had been intent on it having both a garden and a laboratory but found nothing he could afford. The best he could find was a fourth-story apartment on the rue de l'Université, too small for any chemistry equipment. It was another opportunity for him to regret losing all the money he had earned at Laugier, the "stupid mistake" as he described it to Gerhardt.[32]

By October things were in a miserable state. His son, four-year-old Hermann, had fallen and was confined to bed, unable to walk. Laurent's funds had run out, and he "had no more money in his pocket than a rat carries on the end of its tail."[33] He quarreled with the rector of the University of Bordeaux, who had refused to give him any more leave and demanded he return to Bordeaux to teach. Laurent would not. "I would rather sweep the streets here," he wrote, "than return to their damned city!"[34] They stopped paying him, and now even his insufficient half-salary was gone.

In defeat, he returned with his family to live with his in-laws in Luxembourg in mid-October, intending to take any job that would pay him.[35] The news from Paris was a further dagger in his heart. When he had become a corresponding member of the Academy of Sciences in August, it had seemed the answer to his prayers. His work would finally be read before the assembled members of the Academy, and they could judge it for themselves, without having it pass through Dumas or Liebig. He had submitted a paper putting forward his best argument to show the importance of a molecule's arrangement. Chemistry, he explained, was like a game of chess. What mattered was the position of each piece upon the board. Yet most analytical chemists were the equivalent of chess players who violently swept all the pieces off the board, mixed them together, and then sorted them out into separate piles. Their program of analysis, after all, first broke a molecule down into constituent atoms, then counted the atoms to come up with what he called the "brute formula," destroying the very thing they were trying to understand.[36] Laurent instead called for a better approach that would study the arrangement itself—for example, through the analogy of its crystalline form.

But no one heard his argument. The permanent secretary of the Academy, François Arago, had presented the paper, and read out loud only the few snatches that most amused him. The whole affair, he implied, had an air of the ridiculous, beginning with the title, which identified the objects of Laurent's research as phtalamic, oenanthic, and pimaric acids. Laurent had made up these names as part of his own rational, comprehensive system, but they sounded like nonsense to the audience. Laurent had also opened with a rhetorical flourish, his first line declaring chemistry was "the study of things that do not exist."[37] He went on to explain himself, describing how chemists could only measure the constituent atoms of a molecule after they had destroyed the molecule itself in the process of analysis.

But Arago read only the dramatic first line, then skipped all of the explanation and went to the last line, where Laurent pled with his readers, "This is not the moment to fully explain myself on this subject. I desire only to show that, if we want to learn something about the arrangement of atoms, it is indispensable to abandon the path that we have been following."[38] Arago was a masterful showman, with a well-honed sense of comic timing. He paused for effect after reading the line, then announced in the driest of voices that all of that had been, "*très tranchante,*" or "*very incisive.*" The room then exploded in "ironic hilarity," drowning Laurent's most treasured ideas in laughter.[39]

Laurent had predicted people would laugh at his idiosyncratic names, but this hurt more than he had guessed. He felt himself drawn to a precipice, discouraged and unable to see a way forward. There were moments when all he wanted to do was "throw everything to the devil" and never think about chemistry again. Other times, he vowed to return to Paris, find a laboratory, and "grind away like a galley-slave to make everyone recognize the truth."[40] In the end, he could not quit chemistry. "I always come back to my loves," he told Gerhardt. By the end of November, he was heading back to Paris, more determined than ever to show them all.

He was able to find a bit of space in a secret backroom laboratory run by Jérôme Balard, who had just gotten an appointment at the École

Normale (an appointment Laurent, in more hopeful days, had dreamed of getting himself). As a *maître de conferences,* or senior lecturer, Balard did not officially have the right to a lab, but he had gotten around the prohibition by designating a few rooms a "collections depot," inviting his students to do research there and even bringing in his own cot so he could sleep there at night.[41] Laurent began working there in 1846. He took up little space and brought his own instruments. The "extreme delicateness of his feelings" made him embarrassed to be working in someone else's lab, and he limited himself to working with the cheapest possible materials.

Most of the other people in the lab ignored him. He was working on a project to show that substances of the same crystalline form have the same optical activity.[42] Balard had no interest at all in either crystals or optical activity, and his students tended to follow suit. Except for one. He had hired a young assistant, Louis Pasteur, who was intrigued by the "strange, delicate-looking man" in the corner, noting his intense gaze and air of a poet.[43] After Pasteur expressed interest in what he was up to, Laurent carefully walked him through the steps of how to make crystals and measure their angles and optical activity. Pasteur had been doing research to earn a doctorate in chemistry, but now he began to rethink his project and changed it to work on crystals, like Laurent. He was quite taken with his new lab partner, writing to a friend, "M. Laurent is destined to occupy the premier position among chemists in a few years."[44] But these were the words of a naïve boy who had no idea how many enemies Laurent was making.

He was now thoroughly back on Liebig's bad side. He and Gerhardt had gone ahead with their project of starting a journal, which they called the *Comptes rendus des travaux chimiques.* The first issues appeared in 1846, although it was a constant struggle to put out, and they had to switch publishers several times. It had not attracted many followers to their "school" and seemed to have the primary result of making it even harder for them to get anything into Liebig's *Annalen* and Dumas's *Annales.* Liebig raged at their impertinence, calling Laurent and Gerhardt "two self-serving cocks spreading their wings, prancing about the top of a manure pile."[45] He called Gerhardt a highway robber and Laurent a counterfeiter in a series of insults that left Laurent reeling at the injustice of it all. He, Laurent, had nothing: no money, no lab, no reputation, no support. Liebig had everything. He sat, in Laurent's words, "covered in honor, gorged on riches." He did not even work on organic chemistry anymore; he was so busy with his various money-making projects, such as nitrogen fertilizer, food substitutes, and his latest idea, a "meat extract" that concentrated all the nutrients of meat into a kind of drinkable tea. But he had dropped all of that to devote his energy to crushing Laurent and Gerhardt's nascent efforts to make a space for their own ideas. Why, Laurent wrote to him demanding to know, did he, who had all that he wanted, lower himself "to play against me the role of vile slanderer?"[46]

Laurent was trying to support himself teaching private lessons at this point, offering a course on crystallography and the use of the blowpipe in chemical analysis. He was, without question, France's premier expert on the subject. But his meticulously prepared lessons, each an hour and forty-five minutes, only yielded five francs apiece. With just a handful of students, it was a recipe for starvation. Desperate, he sold whatever he could. "Misery has made me commit a crime," he wrote to Gerhardt, admitting that he had sold his course notes to a publisher for 500 francs.[47] The only offer he got was from the Rousseau brothers, who sold chemistry sets to students and wanted to include it as an instruction booklet packaged together in a case with the necessary instruments and reagents.[48] Ashamed of the transaction, Laurent refused to let them include his name as the author. Laurent mused that his life could not continue as it had been going. "What misery! What a dog of a life! Not a penny in the pocket!" He drew a picture of himself plunging headfirst into the Seine, the only conclusion he could see for his impossible situation.[49]

Laurent did his best to maintain his fragile truce with Dumas, but the relationship was fraught. In one letter, Laurent had sketched a physiognomy of Dumas with a two-faced mask, one side scowling, one side smiling, and he never quite knew which side he would get.[50] In the spring

FIGURE 28. Auguste Laurent included this picture of himself jumping headfirst into the Seine in a letter, after the line "In a few months from now, I will have to plunge . . ."

of 1847, Dumas seemed to offer the smiling side, extending a lifeline to the struggling Laurent by allowing him to substitute for some of his courses at the Sorbonne. Dumas gave him strict orders, however, to stick to empirical results and refrain from teaching any of his own idiosyncratic ideas. But Laurent just could not help himself. His lessons gave a vision of chemistry vastly different from that of Dumas. Pasteur was in the audience, writing home that "Laurent's lectures are as bold as his writings," and that he had made "a great sensation amongst chemists."[51] A "sensation" was exactly what Dumas had hoped to avoid. He accused Laurent of espousing theories. "It's not true," Laurent protested—he was simply avoiding all theories, including Dumas's old ones.[52] Dumas stripped Laurent of his duties and suggested that if Laurent wanted to continue working in chemistry, there was a project available in Algeria that he should look into. "They're kicking me out of Paris," he wrote, seeing clearly Dumas's hint that if he ever wanted to teach chemistry

FIGURE 29. Auguste Laurent also drew some caricatures in his letter, including Jean-Baptiste Dumas (*left*) holding a two-faced mask, and a self-portrait (*right*).

again, he would have to go as far as Africa to do it.[53] His career seemed utterly, truly over.

PHILADELPHIA, 1848

At least someone was getting rich after working in Laugier's lab. Eugène Roussel, who had replaced Édouard and Laurent as the director of the laboratory, had been having a very good year. He had lasted only two years at the rue Bourg-l'Abbé, leaving in 1838 to seek his fortune in America. Tucked in his luggage as he crossed the Atlantic was a recipe book of perfumes and liqueurs, favorites of Laugier Père et Fils, which became the basis for his own business. Now, ten years later, he was the most decorated perfumer in America, having just won the only gold medal of his category at the national exhibition in Boston. His booth in Faneuil Hall was a crowd favorite, displaying two enormous busts carved out of soap—one of George Washington, the other of Benjamin Franklin—to honor his new adopted country, where he had just announced his intention to become a citizen. The jury pronounced them "the most remarkable saponaceous productions we have ever examined."[54]

He was based in Philadelphia, where he had set up shop at 75 Chestnut Street when he first arrived. He ran an advertisement in the papers announcing that he had opened an establishment "for the sale of French and English Perfumery."[55] He claimed to have specialties from all the major houses, including Eau de Laugier, and Laugier's special Savon à la Neige. He sold Eau Lustrale from Guerlain and Mint water from Lubin. From England, there was Rowlands genuine Macassar Oil, Rigge's Emollient Vegetable Soap, and Tricopherous or Medicated Compound, for the Growth and Beautifying of the Hair.

In fact, it was a bit of a scam. He made the perfumes, soaps, and oils himself, and ordered counterfeit labels of all the major houses from a printer, M. Pignatel, in Paris. Henri Renaud, who had taken over Laugier Père et Fils, supplied him, being Roussel's primary contact in Paris. When the police raided Renaud's establishment in 1839, he claimed he had no idea how the fake labels got in the shipments he was preparing for Roussel.[56] The Tribunal found him guilty nonetheless, and both he and Pignatel had to pay 250 francs each in damages to Piver, Chardin et Houbigant, and Labouché et Lubin. It was a shameful downturn for such an esteemed house, and Renaud soon left the business, turning Laugier Père et Fils over to Joseph Sichel-Javal, who tried to re-establish its reputation.[57]

Roussel bounced back quickly. The United States did little to regulate this kind of fraud, and a few counterfeit labels were not the worst thing anyone had seen. Before long, he expanded his business, moving his perfume trade to a larger building on 114 Chestnut Street. His reputation for high-quality wares grew alongside the burgeoning market for "Foreign perfumery and Fancy articles," as his advertisements proclaimed. He was the only perfumer of French origin at the exhibitions he attended, and the

FIGURE 30. Eugène Roussel left Laugier Père et Fils to set up this perfumery in Philadelphia, where he pioneered the bottling of carbonated soft drinks.

jury noted his products "far excelled all others of the kind, in this or the former Exhibitions."[58]

But all of his success in perfumery was eclipsed by the historic event taking place at his second location on Prune Street: the invention of bottled soda.[59] The inspiration can be traced back to the recipe book he had brought with him from France, where there were a number of flavored drinks alongside the perfumes, such as Tears of the Widows of Malabar, which was flavored with cloves, mace, and cinnamon, and tinted brown with caramel. Or "The Sighs of Love," which was flavored with rose water and given a rosy pink hue by adding ground cochineal.[60] These recipes were originally intended for liqueurs, but Roussel, arriving in Philadelphia at the same time a vigorous temperance campaign kicked off, found better luck adding the flavors to carbonated water. This had recently become popular in France. Naturally carbonated water from mineral springs had long been considered a healthful cure. The revised national *Pharmacopoeia* of 1837 gave a recipe for carbonating water, by dissolving five parts carbon dioxide in one part water, and suggested adding some lemon sirop to make "limonade."[61] These drinks were not unheard of in the United States, occasionally sold directly from the fountain. But Roussel was the first to successfully bottle them, capturing the fizzy water with a tight cork that left off a satisfying pop when opened. Soon after he opened his store, he provided the local newspaper with a few samples. They reported favorably on his "artificial mineral waters, put up on the French plan."[62]

His second store on Prune Street become the epicenter of a new craze—"mineral water fever," as the papers put it.[63] When summer temperatures hit 95° F in the shade, he sent bottles of flavored soda water to major newspapers, accompanied by a shipment of ice. "Public benefactors," said the *Evening Argus* of Roussel and his operation, "They dispense their favors with a liberal hand, and thirsty, heated, panting human nature rejoices." The delightful beverage, they insisted, was "not only palatable and cooling, but it is a positive health promoter." Roussel advertised the recommendation of the medical profession, touting its "gently stimulating and wholesome qualities," which gave a "healthy tone to the Stomach and Bowels."[64] He went from selling about 100 bottles a day to well over 1,000, with three bottling machines running full time.[65] "Pop! Pop!" went the headlines of Philadelphia, celebrating the invention of this new "delightful beverage" by the immigrant perfumer. When the temperance movement finally succeeded in passing the Prohibitory Liquor Law in 1855, Roussel stepped in to provide thirsty Philadelphians with his new signature Carbonated Ginger. It became the drink of the summer, a "pleasant and wholesome beverage" that made a good substitute for "the more fiery potations which recent legislation has put under ban."[66] The newspapers credited his drinks with "the prolongation of life and multiplication of life's enjoyments"—elixirs of vitality for a new age.[67]

11
The Synthetic Age

PARIS, RUE BOURG-L'ABBÉ, 1848

Édouard Laugier was back on the rue Bourg-l'Abbé. He worked as a distiller down the street from the original perfume house, which was now under the ownership of Joseph Sichel-Javal. While Sichel-Javal was trying, as the papers put it, to return the store to "the heights that M. Laugier had been able to bring it," Édouard had moved into no. 54, sharing the building with Mlle. Germond, who sold lingerie.[1]

The neighborhood had changed since he had grown up there as a child, growing both in economic importance and political agitation. Its merchants had grown rich in the new era of industry, but for many that only highlighted their feeling of disenfranchisement. France was still a monarchy, and while it did have an elected legislature, the vote remained tightly restricted. Out of a population of some 35 million, fewer than a quarter million could vote in national elections, keeping politics in the hands of a small elite.[2] The Saint-Simonian vision of technology bringing about a world of true equality was starting to look more and more like a utopian dream. Laugier and Laurent's Left Bank friend, Jean Reynaud, had taken to calling the new era one of "industrial feudalism," where power remained concentrated in a few hands, and hinted that violent insurrection was the only remedy.[3] Calls for social reform had spread from the philosophical discussions of left-bank intellectuals to more urgent demands of an angry middle class, and Laugier's neighborhood was, as Victor Hugo put it, the primary "field of combat" for these bourgeois revolutionaries.[4]

Rue Bourg-l'Abbé had been the site of France's last serious attempt to start a revolution, in 1839. The target was a gun shop, Lepage Frères, located at number 22. Some 1,000 men had converged on the street on 12 May, summoned through underground networks by the revolutionary leaders Armand Barbès and Auguste Blanqui. They did their best to mill about unobtrusively in the nearby shops, although some of the proprietors found it odd that they were so busy on a Sunday, and that the men crowding their stores all seemed to know each other. At 2 p.m., Blanqui and Barbes gave the signal, and everyone stormed the gun shop, emptying it of 310 rifles and 100 pistols.[5] The plan was to head to the Hotel de Ville, while the rest of Paris, inspired by their bravery, rose to fight alongside them. But the reinforcements never came, and their own numbers began to dwindle. As night fell, and the police pressed their counterattack, they retreated back to the rue Bourg-l'Abbé, behind the barricades they had constructed. By the morning, the barricades were subdued, Barbès injured and arrested, and Blanqui disappeared into the night.

Things were quiet for a while in Paris, after the embarrassing fizzle of a failed revolt. But the underlying issues remained, and calls for reform continued. Suppression grew even more severe, as the king banned political meetings, demonstrations, and even, by February 1848, informal gatherings of the political opposition known as "banquets." This last restriction was a step too far, and the day after it was announced, a crowd of thousands of Parisians gathered before the Chamber of Deputies to protest. The police eventually dispersed them, but it was obvious the anger had not dissipated. The authorities prepared for more violence, announcing a 7 p.m. curfew and sending the police to line major boulevards such as Saint-Denis and Saint-Martin. One of the brothers of Lepage Frères, who shared the authorities' concerns, had called for reinforcements, and roughly a dozen police officers had entered the store on rue Bourg-l'Abbé unnoticed, to protect it if needed.

As night fell on 22 February, a group of fifty men attacked Lepage Frères, trying to break through the main carriage doors.[6] These had been reinforced with iron after 1839, and even though various groups kept hammering at them through the night, the best they could do was shift the position of some of the middle panels. Eventually, they created small cracks where they could peek through to the other side—a dark passageway that had a spiral staircase to the left leading to the gun store's entrance and an open courtyard at the back. Someone caught a glimpse of a police officer's uniform and raised the alarm. Realizing they had been recognized, the police began shooting through the cracks in the door, killing one of the insurgents. The crowd scattered, but soon regrouped and continued their attack.

They worked through the night building barricades to fight behind. They turned over a large wagon, formerly used for hauling flour, and positioned it between the rue Bourg-l'Abbé and rue Greneta, filling any gaps with cobblestones, furniture, and whatever else they could get their hands on. Victor Hugo arrived in the morning. He had woken before dawn on 23 February to wander the streets of Paris, noting the dozens of barricades that had gone up the night before, although only the one on rue Bourg-l'Abbé had the designation "good barricade" beside it in his notes.[7] He interviewed the local shopkeepers, who generally supported the action. "They are building the barricades in a friendly way," without annoying the neighborhood, one noted, sympathizing with the fighters who had spent the night camped behind the flour cart, a cloaca of mud after the rain.[8]

As the day broke, the police moved in, firing cannons and rifles.[9] One contingent of police, some fifty strong, stormed the rue Bourg-l'Abbé and took up a defensive position in the passage du Petit Hurleur, directly across from Laugier Père et Fils. They shot indiscriminately into the street; the insurgents dodged in and out of buildings and shot back. Gunfire continued throughout the day. By the afternoon, the corner of the building by the alley was so riddled with bullets that it collapsed into rubble.[10] As the number of insurgents continued to swell into the thousands, the police in the alley retreated to join the others in Lepage's courtyard. They reinforced the door behind them and began to contemplate their situation.

The police decided to negotiate their surrender. After an hour of bargaining, they emerged from the courtyard, stripped of their weapons and the distinctive cylindrical caps, called shakos, that were part of their uniform. They marched single file through the streets, surrounded by throngs carrying torches and weapons and singing the Marseillaise.[11] The crowd retraced the path to the Hôtel de Ville that Barbès and Blanqui had taken in 1839, only this time, their numbers swelled as they went. People all over Paris had risen up—barricades riddled the city. The insurrectionists deposited their police prisoners safely at the Hôtel de Ville, but the weakness of the king's position had become evident. He soon abdicated and fled to England as "John Smith," the last king that France would ever have.

France was a republic once again, with the radicals and socialists now in charge. Barbès went straight from prison to a seat on the constituent assembly, where he was joined by Blanqui, Pierre Leroux, and Jean Reynaud. Their friend, the former Saint-Simonian Hippolyte Carnot, became the minister of education, and he brought in Reynaud as the deputy Secretary of State.[12] It was a compilation of Laurent's oldest friends and political fellow travelers, and as it happened, they needed a chemist.[13] The Paris Mint had employed the chemist Jean d'Arcet for decades to assay coins and oversee production, and the position had sat open since his death in 1844. Within a month of the revolution, on March 13, 1848, it was Laurent's.

Laurent finally had what he always wanted: a position in Paris. *La Monnaie*, the Paris Mint, was a venerable institution, dating back over a thousand years and housed in a palatial building on the banks of the Seine, right off of Pont Neuf. Laurent, however, did not work in the airy, neoclassical building itself, but down in the cellar, in an underground room with no windows, constantly damp from the river water seeping through the walls. It was unfurnished, as well, and he had to bring in his own equipment. He moved in anyway, determined to make it work. He had waited far too long for this opportunity to arise.

FIGURE 31. The barricade on rue Saint-Martin, near the rue Bourg-l'Abbé, where police were captured trying to defend its gun shop.

PARIS, RUE D'ULM, 1848

Louis Pasteur was working in Jérôme Balard's laboratory when the revolution broke out in February. He could hear the gunshots from the barricades and was thrilled by the fight for liberty and justice. "It is a great and a sublime doctrine that is now being unfolded before our eyes," he wrote home, "and if it were necessary, I would fight heartily for the holy cause of the Republic."[14] He joined the National Guard, which had emerged in the wake of the revolution as the primary defenders of the Republic. "What a transformation of our whole being!" he exclaimed, seeing France born anew as the land of free men.[15] One day in April, while crossing the Place du Panthéon, he saw a wooden structure with "Altar of Our Country" written on it. Someone explained that it was for accepting donations to the Republic, so Pasteur went back to his lodgings at the École Normale, emptied his drawers, and returned to place the entirety of his savings—150 francs—on the altar.

Nor was his excitement for his work in the chemistry lab dampened by this revolutionary upheaval. If anything, it paralleled the sense of a world being born anew. He had completed his dissertation the previous fall and was now engaged in a new project that probed the very contours of life. A notebook lay open on his workbench, the phrase "Tartrates (Questions to resolve)" written across the top.[16] The substance in question was simple enough, being a common by-product of making wine. Vintners usually found tartar in the bottom of the barrels after emptying them, and had taken to scraping it out, grinding it up, and purifying it to sell as cream of tartar, a kitchen staple used for making baking powder. But anyone who opened a bottle of wine could see tartar, accreted in large crystals on the underside of the cork. For red wines, the crystals could take on a deep, amethyst-like purple. For white wines, they were clear, sparkling like diamonds. Indeed, "wine diamonds" was the name used by oenophiles.

Pasteur had, like Laurent, grown up among the vineyards of Eastern France. His two closest friends, Jules and Altin Vercel, were the sons of the neighboring vintner, and he spent his days playing there and helping with the annual harvest.[17] His own family had run a tannery, and they collected the tartrates from the spent mash to treat leather. At fifteen, he was sent to Paris to study, together with the vintner's son, Jules. But the shock was too great. His missed his home, the contour of the land, its smells. "If I could only get a whiff of the tannery yard," he confided in Jules, "I feel I should be cured."[18] He lasted scarcely a month.

He tried again at the age of eighteen, finding lodging in Paris above a public bathhouse. His parents wrote him solicitous letters encouraging healthy habits. "Take some nice baths," his mother wrote, although she insisted they should not be too hot, as his heavy workload might lead to overheating. His father suggested he put some drops of cologne water in his hair to prevent headaches and strengthen his eyesight.[19] He also told him that he should avoid running around Paris late at night, and instead of going out to see plays, he would do better to stay home and read them himself, and would likely come away with a better understanding of them, anyway.[20]

One of the first things Pasteur did in Paris, before his own classes even started, was to sit in on a lecture by Jean-Baptiste Dumas at the Sorbonne. "You cannot believe how many people there are at this course," he wrote home.[21] Some 600 or 700 people filled the immense room, and you needed to get there half an hour early to get a good seat. He compared it to an evening at the theater, complete with thunderous applause at the end. Dumas's soaring rhetoric, he explained, "could set fire to the soul," and appears to have lit something in Pasteur, who was soon accepted in the sciences section of the École Normale, where he haunted the chemistry labs every chance he got. When he graduated in 1846, he wrote to Dumas asking if he could work as his teaching assistant at the École Centrale. The answer was no, but Pasteur was hired by Balard, who needed an assistant for his clandestine lab at the École Normale, the one where Pasteur would meet Laurent.

Pasteur never forgot the moment Laurent first called him over to peer into his microscope at some sodium tungstate crystals. It had looked perfectly pure at first glance, but Laurent showed him that if you looked at the structures of the crystals, you could divide them into three distinct kinds. The two men shared an artistic bent. Pasteur had even thought of becoming an artist before discovering science, and they both continued to draw constantly in their lab notebooks—caricatures and grotesque physiognomies. Perhaps that is why they liked to visualize the arrangements of atoms and think about how a molecule's structure might look, when so much of the chemistry community dismissed this as dangerous speculation. In any case, Pasteur was hooked, and Laurent began teaching him how to work with crystals. He had been working on his thesis to earn a doctorate in chemistry, but he now abandoned his old project and began a new one with Laurent.[22] Pasteur kept his new activity largely under wraps, writing only to a friend to convey his excitement but urging him not to tell anybody. In the end, Pasteur submitted two separate theses in

August 1847. One was in chemistry, which consisted of a straightforward chemical analysis of arsenious acid and its salts, confirmed with crystallographic methods. The other was in physics, and made precisely the point that Laurent had been arguing to the chemical community, to an indifferent response: substances of the same crystalline form had the same optical activity.[23] It was a powerful claim that pointed, both Laurent and Pasteur were convinced, to some aspect of the molecule's physical organization. But these were not the kinds of ideas by which someone earned a doctorate in chemistry at the time.

Now, in the spring of 1848, doubly doctored and with the tocsin of liberty sounding around him, Pasteur turned to the heart of the question: was there something special about the organization of the molecules of living things? Could a study of tartrates hold the key? Chaptal, in his study of wine, had seen a close congruence between tartrates and the "vegeto-animal principle" that he proposed as a vital force particular to living things. Biot had been virtually obsessed with tartrates, studying them for decades in his effort to use optical activity to distinguish between the living and non-living. And now there was another wrinkle to the story. In addition to the "natural tartrate" that formed in the process of fermentation, when acid in the grapes reacted with a potassium salt, there was also an "artificial tartrate" created as a by-product of industrial production. An industrialist, Karl Kestner, had first noticed this substance as a waste product of his chemical factory in Thann, Alsace, in the 1820s.[24] He showed it to Gay-Lussac when the famous chemist came for a visit, who gave it the name racemic acid. As analytical techniques improved in the 1830s, chemists noticed that it had the exact same ratios of carbon, oxygen, and hydrogen as tartaric acid, prompting Berzelius to rename it paratartaric acid, to indicate it was an isomer of tartaric acid. By this time there were several cases of isomers that had the same number of carbons and hydrogens. Isomers could usually be distinguished by different properties. Berzelius, hoping to find differences, asked Mitscherlich, at the University of Berlin, to prepare salts of the compounds and compare the resulting crystals.

Mitscherlich had, like Laurent, published his first scientific paper on mineralogy, and he was one of the few other chemists who was skilled with a goniometer. He put it to use now, comparing the salts made with grape tartrates with some made from paratartrates he got from a factory in Saxon. Not finding any differences, he put the work aside for several years, until he heard of Biot's demonstration that tartaric acid was optically active and racemic acid was not. He repeated his experiments, once again finding a remarkable identity: they had the same crystalline form, with the same angles, the same specific weight, the same double refraction, and the same angles between their optic axes. Dissolved in water, their refraction was the same.[25] One difference alone distinguished them: the natural substance was optically active and the artificial substance was not. Mitscherlich sent a note of his findings to Biot, who pronounced his experiments "beautiful" and his results "curious."[26] Mitscherlich had provided him with samples of his material, which Biot investigated relentlessly, hoping to uncover some clue about the organization of living bodies.

This was the challenge that Pasteur picked up. He was able to borrow a bit of the paratartaric tartrate Mitscherlich had sent to Paris, and then made crystals of both it and some normal tartrate by reacting them with potassium, ammonia, or the like to produce salts. He looked at the normal tartrates first. At first glance, the crystals they formed seemed perfectly symmetric, but after a while he noticed something entirely new: what he called "little faces that betrayed its asymmetry," that is, tiny protrusions that jutted out of one side but not the other.[27] There was a word for the asymmetry that Pasteur was seeing: hemihedral. This type of asymmetry was well known among crystallographers and usually associated with the crystal quartz. This seemed to Pasteur an important clue. Quartz was an inorganic mineral that was nonetheless optically active. Although it was symmetric at a molecular level, it crystallized into an asymmetric form, and this seemed to be what "pulled" the light to one side in optical activity. Pasteur thought a similar process might be responsible for the optical activity of the tartaric acid.

His next step was to look at the crystals of the "artificial" paratartrates. Since these were not optically active, he thought they would perhaps be perfectly symmetrical, thus finally finding the distinguishing feature Mitscherlich had missed. But what he actually found was much stranger. They, too, were hemihedral, but unlike the tartrates, whose faces were all on one side, the paratartrates had some faces to the left and others to the right. When he first observed this phenomenon, he recounted, "my heart skipped a beat."[28] He felt on the edge of solving the mystery. Slowly, painstakingly, he separated out the two groups by hand: one whose crystalline structure was identical to the plant-derived tartrates and one whose crystals were mirror images of those. He tested each batch in a polarimeter and found that they had now become optically active. The one identical to the tartrates deviated the plane of polarization about 7 degrees to the right, while the mirror-image batch deviated it the same amount to the left, which explained why they cancelled each other out and looked inactive when combined.[29] Pasteur ran out of his room like Archimedes, threw his arms around the first person he saw, and dragged him to the Luxembourg Garden, explaining his discovery along the way.[30]

He took his news to Balard, who would be able to report it at the Academy of Sciences for him. Balard did not even wait for an official session but instead gossiped loudly about it in the corner of the library where academics came to socialize. Dumas was present, and listened seriously. Biot overheard and approached Balard, asking, "Are you quite sure of it?"[31] He wanted to meet the young man who had solved the problem he could not, and investigate his results. That was how Pasteur came to be knocking on Biot's door at the Collège de France one morning in the early spring of 1848. Pasteur reported that he sensed suspicion in the old man's voice and gestures.[32] Biot demanded that Pasteur prepare the paratartrate salt in front in him, supplying him materials from his own stores—soda, ammonia, and most crucially, some of the racemic acid he had himself been working with. He watched over Pasteur as he went through the manipulations, then placed the liquid in a quiet corner of his apartment, where it could crystallize undisturbed, and told Pasteur he would send for him when the crystals were ready.

Several days later, summoned again, Pasteur sorted the crystals into the two types, hunting down the best specimens, wiping off the *eau-mere* that adhered to them. This experiment has proved devilishly difficult for the many who have tried to repeat it. The difference between the two crystals is normally hard to see, and it seems Pasteur's dogged care resulted in particularly large crystals. There was also some luck involved, when Pasteur cooled his tartaric acid on the windowsill. The left- and right-handed crystals only separate below 26° C and would not have revealed themselves in a warmer month. But he was able to get them again for Biot, and at this point, the older man took over. He prepared the salts in a solution that he placed in a polarimeter. When he verified what Pasteur had seen, he took him by the arm in delight and proclaimed, "My dear boy, I have loved science so much during my life, that this touches my very heart."[33] After working on the question for thirty years, he finally had an answer. Natural products were optically active because they possessed a fundamental asymmetry that artificial ones did not. And it further entrenched his suspicion that it was folly to try to replicate these natural substances by artificial means.

LONDON, 1851

August Hofmann was by now a celebrated fixture of the London social scene. He gave chemistry lectures in Windsor Castle before Queen Victoria herself and dined frequently with the British aristocracy, where he was so friendly, agreeable, and funny that people almost forgot he was German. His very appearance transformed, as the tiny, pointed goatee of his youth grew into the long bushy beard so characteristic of the Victorian age. And he had grown wealthy through his work as a consultant to industry. The Royal College of Chemistry, which he directed, was far more oriented to practical applications than the universities of France or Germany, and many of the students working under him had industrial ties.

He himself embraced the role, comparing his position as director to "an industrialist in command of a splendid machine."[34]

Hofmann had long ago left behind any allegiance to Laurent. When he visited Paris in 1851, Thénard gave a dinner for him at his house in Fontenay-aux-Roses, with bottle after bottle of the finest wines. All the chemists of Paris were there, Thénard recalled in a letter to a friend, with the exception of "THOSE TWO," who went unnamed but were undoubtedly Laurent and Gerhardt.[35] But despite their lack of contact, Hofmann's work still bore the echoes of his work with Laurent. Although he steered well clear of Laurent's theoretical claims, his approach used his idea of a fundamental nucleus, treating the chemical types as building blocks that could be used in construction. Hofmann also continued the dream he had shared with Laurent of making artificial molecules from coal tar. His work in Giessen became the basis for a new era of "synthetic experiments," where chemists not only broke down molecules into their constituent parts but built them back up as well.[36] He still dreamed of making quinine out of coal tar, noting that the first chemist to do it would be celebrated as a benefactor of humanity. "Such a transformation has not yet been accomplished," he admitted, "but that alone does not imply that it is impossible."[37]

Nowhere would the new industrial chemistry get a better showing than at the Great Exhibition of 1851. It had been the pet project of Prince Albert, who asked Hofmann to assist in organizing the exhibition. It was Britain's answer to the series of Great Exhibitions of Products of French Industry that had begun under Napoleon, and their attempt to show that Britain, not France, was winning the race of industrialization. Prince Albert was determined that Britain's version be bigger, grander, and more impressive than anything France had done. He invited exhibitors from all over globe, in part to drive home the point that France had not invited Britain to its exhibitions because it knew Britain could flood the market with products produced in quantity and for less money.

The plan was for a new Versailles for the machine age. When Louis XIV first built Versailles, he wanted to show off the recent technological breakthroughs of French glassmakers. They had just invented a technique at Saint Gobain for rolling large, flat panes of glass. The king had ordered 357 of them, painted on the back with reflective silver, to line his signature Hall of Mirrors. By 1851, machines had automated nearly every step of the process, as steam engines provided the power for pressing, rolling, grinding, and polishing the glass. The exhibition organizers capitalized on this, ordering some 300,000 individual glass panes for the exhibit's main building.[38] The result was an immense building, nearly a million square feet, with walls and roof alike made entirely of glass in an iron frame.

Londoners nicknamed it the Crystal Palace, and inside were over 100,000 exhibits of the wonderous, new, and strange—"a sort of fairyland" that plunged the visitor into a "state of bewilderment," wrote Lewis Carroll, who visited at the age of nineteen.[39] Visitors entering the spacious transom were immediately presented with a perfume fountain, where they could dip their handkerchiefs to scent them. It was the work of Eugène Rimmel, a French-born perfumer who had set up shop in London and now commanded a reputation as Britain's finest perfumer.[40] He was considered a marketing genius, associating his name with fine luxury while also providing more affordable products such as his best-selling item, "toilet vinegar," which was well within middle-class budgets. Whereas Louis XIV's perfume fountain in Versailles had been a potent symbol of extravagant wealth, Rimmel's was available to anyone able to pay the entrance fee—a mere shilling on discounted weekdays.

Most marvelous of all among the perfume displays were the new synthetic products. Collas had a booth in the French section, showing several items that had begun their lives as coal, including not only his Essence of Mirbane and samples of almond-scented soap, but also a brand-new artificial essence: pineapple, a tropical fruit so prized by the English that they displayed them in their homes as status symbols. He had come across the scent by happenstance. The process of rectifying nitrobenzene left behind a residue, which, Collas discovered, after adding some alcohol, produced an ether that gave off the aroma of pineapples, with

FIGURE 32. Eugène Rimmel's perfume fountain at London's 1851 Crystal Palace Exhibition. Visitors could dip their handkerchiefs in to pick up the scent.

maybe a hint of strawberries. Collas had patented it in 1850 and soon began selling it under the name Essence d'Ananas for flavoring sirops, ice creams, and candies. And that was not all. The French section also displayed an artificial wintergreen, using Cahours's process of making it from salicylic acid. The British showed off their own "pear drops"—a hard candy sweetened with barley sugar and flavored with amyl acetate, a chemical that Hofmann admitted did not, by itself, smell great, but if used in minute quantities, could evoke the "agreeable odor of the Jargonelle pear." Similar chemicals evoked the aroma of apples and cognac, while Hofmann noted of several less-successful chemicals that "their resemblance to the aromas specified was very doubtful."[41]

Hofmann was in charge of judging these products. Prince Albert had placed him on the jury for "miscellaneous manufactures and small wares," which covered a range of objects from candles to umbrellas to "apparatus for manly games" such as rackets and archery.[42] But perfume and soap were listed first among these, and the jury noted in their report that they were without doubt the most important of the group, even reproducing Liebig's adage that a country's wealth and civilization could be measured by its soap consumption.[43] The jury members visited the exhibits and spoke with the manufacturers about the extent of their production. Hofmann took home samples to analyze in his lab.

Hofmann predicted that these first, somewhat dubious efforts were merely the beginning of a glorious new age for perfumery. But he also seemed taken aback that the first commercially viable artificial almond scent should appear in the French section. Chatting with Collas at the exhibition, he had been surprised to learn how considerable his production already was. When he analyzed the sample he was given, he found that it was indeed the same nitrobenzene that his student Charles Blachford Mansfield had produced, and the report of the jury emphasized Mansfield's role in inventing the process for producing it cheaply. The report was faint in its praise of Collas, noting that all of the various products he had on display were from nitrobenzene of "more or less purity" and dismissing his use of the "fanciful name" of Essence of Mirbane.[44] The jury awarded him an honorable mention and *medaille de prix,* rather weak accolades for his accomplishments.

But Collas turned his showing at the Great Exhibition into a marketing opportunity, and sales of his Essence of Mirbane picked up back in France. Quesneville showcased the new perfume in the next volume of his ongoing series, *Secrets of the Arts,* where Collas revealed the details of how he made it.[45] Madame Celnart added it to the 1854 edition of her *Manual of the Perfumer,* echoing the claim that it had the advantage over the natural oil of not containing any toxic substances, such as cyanide.[46] As the sales of mirbane soared, its uses multiplied beyond perfume and soap to include candies, pastries, sirops, hair oil, face creams, cleaning pastes, and more. It was particularly prized for getting the wet-dog smell out of wool. People rubbed it on themselves, ate it, drank it, wore it in their clothing, massaged it in their scalps, and cleaned their homes with it. Some may have reported headaches or faintness, but that could be from anything, couldn't it?

EICH, LUXEMBOURG, 1851

And what chemicals had Laurent ingested over the years, and to what effect? It was a question that he asked himself over the spring of 1851, as he took his daily rest in the sun and tried, with diminishing success, to fill his lungs with fresh air. He was dying of consumption, the fate of so many Romantic poets and artists. Now known as tuberculosis and understood to be a bacterial infection, physicians at the time had debated its cause and often attributed it to an imbalance of the passions, appearing particularly in those whose strong emotions outstripped their physical bodies.[47] Laurent had indeed led a life of intense pursuit, and it seemed a particularly cruel irony that once he had achieved the object of this desire—a position in Paris, a laboratory of his own at the Mint—he should become too sick to use it. He had hardly begun working at the Mint when he started coughing up blood, the telltale sign of the disease. Death came slowly, over the course of years, as his body wasted away as if being eaten from the inside. It was unrelenting, and Laurent had no illusions that he would make it out of his forties.

Everything had unraveled so quickly: his health, his work, the new republic that was supposed to usher in an era of justice and equality. It had felt like the dawn of a new era in 1848, when France declared itself a Universal Republic, the only one in existence that allowed all adult men to vote, with no property requirements. But the first presidential election was a disappointment. Without much of a civics education, the newly franchised peasantry voted for the person with the most recognizable name: the notoriously dim Louis-Napoleon Bonaparte, whose only qualification was being the original Napoleon's nephew. It was part of a conservative resurgence that also swept Dumas back into power, electing him to the National Legislative Assembly. He and Louis-Napoleon got along just fine. "You will be my Chaptal" the new president declared, and appointed Dumas minister of agriculture and commerce in October 1849.[48]

By this time, Laurent's doctors had forbidden him from working in the damp cellars of the Mint. There had been a glimmer of hope, in 1850, when the chair of chemistry at the Collège de France became available. Pelouze had taken over the position when Thénard retired in 1845, but now he, too, was leaving. Biot was the senior physics professor at the Collège de France, and he rallied the other faculty to support Laurent for the chair. Unfortunately the decision was not up to them but placed in the hands of a secret committee drawn from the chemical section of the Academy of Sciences. This was a group dominated by Dumas, who preferred another candidate: Balard. It seemed yet another cruel lesson for Laurent in how unequally the resources of the world were divided. Balard not only had a perfectly good laboratory at the École Normale, where Laurent had first met Pasteur, but he had been given another, even better, one at the faculty of sciences when he became the (second) chair of chemistry there in 1847. The Collège de France would be his third major laboratory, while Laurent had a moldy underground burrow, completely unusable. Biot was so outraged that he commandeered the Academy floor to deliver an angry diatribe on the issue of "scientific justice."[49] Balard already had all the means to undertake any work he was inspired to do, which, Biot pointed out, was essentially nothing. Laurent, on the other hand, had produced paper after paper solely "by his unaided personal efforts at the cost of heavy sacrifices."[50] Giving the lab to Balard would add nothing, but it could be a lifeline to Laurent. It was like heaping food on the plate of a full man while denying it to one starving beside him. Aiming his moral shaming directly at Dumas and the Academy's chemical section, Biot had his speech printed and distributed it himself. But his efforts were to no avail: Balard got the job.

Laurent's friends abandoned him one by one. Hofmann had stopped all communication. Gerhardt had pulled away, recognizing that "that poor Laurent," as he called him, could hardly function as an ally.[51] Even Pasteur distanced himself. He was trying to publish his doctoral thesis, and to do so, he had to go back through and erase any acknowledgment of his debt to Laurent.[52] His own politics had changed; he had replaced his youthful enthusiasm for the revolution of 1848 with a more conservative desire for order. (His mother had died of an aneurysm in May of 1848, and he blamed the chaos of the revolution for it.) He had begun to align himself with Dumas and, now savvier in the politics of Paris chemistry, cemented the relationship by repudiating Laurent. He wrote a letter of self-criticism to Dumas explaining that he had been young and impressionable when he first met Laurent, and he allowed himself to get swept up in "hypotheses without basis" before being "quickly enlightened by your advice."[53]

Laurent and his family, which now included a new baby girl, moved back to Eich early in 1851. They lived with his in-laws, now thankful that the house stood outside the city gates, surrounded by trees and ample gardens. Fresh air and sunlight were the only known treatment for consumption, and patients often spent months or years in sanatoriums, passing their days reclining outside in regimented repose. Laurent did the same at his in-laws, with obligations only to "eat, rest, go for walks, yawn."[54] His great solace was watching the buds appear on the trees and the flowers take bloom. He had grown up in the countryside, attentive to its seasonal rhythms, and it occurred to him that his life would have been better if he had never left it. Instead he had read Biot's textbook on experimental physics as a schoolboy and become entranced by the prospect of an orderly, explicable world. He wrote to Biot now, admitting that there were times, "seeing all these trees and flowers, all these plants coming out of the ground," when he cursed the older physicist for setting him on the path of science in the first place, and drawing him away from the vineyards of his childhood. "When I think that I passed the best years of my life miserably confined in dark, damp, infected laboratories, I wonder if I was not suffering from mental alienation . . . But science is such a beautiful thing! You know, if I had a nice laboratory open on this beautiful garden I believe that, like the opium eaters, I would never want to lose my madness."[55]

He had a vision for the job he wanted: a chemist that did not work in a lab at all but stayed outdoors, investigating organic substances while they were still alive, growing and taking form. He wrote up a proposal for Dumas, now minister of agriculture, suggesting that they would together create an institution for studying vegetable physiology. Laurent could travel through France, studying plants on its sunny hillsides, and produce useful contributions such as a map of the soil of France suited to growing particular crops. It was an ambitious plan for someone so sick, but Laurent felt himself in the grips of a terrible predicament. If he died now, his family would be left with nothing. Normally, a widow could draw upon her husband's pension, but he had none. He had given up any claims to retirement when he left his position in Bordeaux. And he wouldn't get any retirement payments from the Mint unless he stayed there until he was sixty, still sixteen years away. But that was impossible. There was a new ministerial ordinance that required him to be present in his laboratory from 9:30 am to 4:00 pm, a requirement that would literally kill him. "I've made a bad calculation," he admitted.[56]

Laurent put the proposal in Biot's hands and begged him to petition Dumas. Biot was, by this time, the only contact Laurent had left among elite scientific circles, and the two shared a correspondence back and forth that spoke frankly of Laurent's impending death. Biot himself had buried his wife and only child in the preceding year. His wife, Gabrielle, had been with him fifty-four years, supporting his work and translating German and English papers for him. His son, Édouard, who was about Laurent's age, had been a respected sinologist who was working on a translation of some ancient Chinese ceremonial rites when he died. The father, approaching eighty, learned enough Chinese to see the work through to completion after his son's death.[57] Now he took Laurent under his wing, approaching Dumas with his friend's proposal. The answer, however, was no.

By the time Biot wrote back, Laurent was no longer in Eich but had gone to Hyères, on the southernmost tip of the French coast. Located between Saint-Tropez and Marseille, it had recently become popular with consumptives hoping to be cured by its sun and pure air. Laurent spent his days in the sun, although he admitted that even that did not seem to be helping much. His response to Biot focused on making arrangements for after his death and asked for help placing his son, eleven-year-old Hermann. He was also, he let on, working furiously on a major treatise: a textbook that laid out a rational classification of chemicals based on molecular arrangement. It was a race against time to get down his own, unique vision of chemistry. But not even his changed locale could help him win it, as he soon slipped into a delirium, hallucinating in the end that he was back in his laboratory, tending his experiments.

His death on 15 April 1853 came exactly as he feared, with his textbook still unfinished, and his wife and two children left destitute. His funeral was a sad, hastily arranged affair, with only twenty or thirty people attending and no words spoken at the grave.[58] Biot tried to fulfill his promises, pressing his colleagues to give money for Laurent's family. They pulled together a donation of 20,000 francs, with many expressing regret at how badly he had been treated.[59] Biot also petitioned the minister of public instruction to grant his widow a pension, and they agreed to provide her 1,000 francs a year.[60] Francine lived an additional sixty-one years, dying only weeks before the outbreak of World War I. Dumas proved particularly contrite after Laurent's death, supporting Biot's efforts with a testimony that emphasized Laurent's "pure and honorable life" and his "elevated spirit."[61] Dumas took special interest in the son, Hermann, helping to secure him a place in a good school. Hermann went on to become a mathematician and obtained what had remained forever out of his father's grasp: a good job in Paris, at the École Polytechnique.

With Laurent's family taken care of, Biot turned to Laurent's unfinished manuscript, gathering the pages to see if anything could be done with them. At least this time he would not have to learn Chinese.

PARIS, RUE BOURG L'ABBÉ, 1853

From the corner of rue Greneta, where he now worked, Édouard Laugier could watch the demolition of the house where he had grown up, and the perfume shop that had occupied its first floor. The entire street was being razed to the ground, with every last cobblestone torn up and carted away. It was, everyone knew, a punishment for the revolution, and a guarantee that no one would ever build barricades there again. Louis-Napoleon had ordered a new boulevard, massive and wide, to be built over the top of where the rue Bourg-l'Abbé had once been. Named the Boulevard du Centre, it was to be the primary north / south axis through the center of Paris, giving access to the Gare du Nord, the recently completed railroad station. "Le vieux Paris n'est plus," wrote Baudelaire. Old Paris was no more, and this included the perfume house of Laugier Père et Fils.

Louis-Napoleon had placed the self-described "demolition artist" Georges-Eugène Haussmann in charge of remaking the map of Paris.[62] Haussmann reported that on his first day in office, Napoleon presented him with a map of Paris, marked in red where the new streets were to be constructed.[63] The plan cut wide swathes through the city's oldest neighborhoods, all in the name, so he said, of hygiene and cleanliness. The dark, narrow streets trapped unhealthy miasmas and kept out the sun's healthy rays. Underneath his new boulevards, he built a network of large sewers—his proudest accomplishment that soon became something of a tourist attraction.[64] He cast his efforts as an all-out war against the bad smells of Paris. Haussmann took full advantage of a new law of expropriation, allowing him to seize property as needed. Édouard was merely one of an estimated 350,000 people he displaced.[65]

But the real goal of the plan was to consolidate power and create wide spaces for his troops to march with cannons. "It's the evisceration of old Paris," Haussmann said, "of a neighbourhood of riots and barricades."[66] Louis-Napoleon had abandoned the pretense of democratic elections, declaring himself the Emperor Napoleon III in a move taken from the first Napoleon, prompting an appalled Karl Marx to remark that history does indeed repeat itself, "the first time as tragedy, the second time as farce."[67]

Nicknamed "Napoleon le petit," the nephew was in every way a pale imitation of his uncle, down to his morning *toilette*. Napoleon III, too, liked to begin his mornings with a bath and a cup of tea. But where his uncle always took his with a healthy dose of Eau de Cologne, the nephew took his plain and unadulterated. Perhaps it was a holdover from his childhood, when his father wrote out the following list of instructions: "My son is to wash his feet once a week, clean his nails with lemon, his hands with bran, and never with soap. He must not use Eau de Cologne or any other perfume."[68] Perhaps it was part of the bigger trend in which men were wearing less perfume and leaving the floral scents to women.

It was a paradoxical time for perfume. The market was bigger than ever. But it was also finalizing its divorce with its medical origins. The fine line between perfume and medicinal elixir had grown sharper, with people rarely drinking perfume anymore. The division was soon enshrined in the tax code through the process of denaturation—in which perfumers added poisonous methanol, or the spirit of wood, to make their product undrinkable, so they could be taxed at a lower rate. Britain started adding methanol in 1855, France in 1872.[69] Gone were the days when perfumes could double as healing elixirs, harnessing the vital power of plants. The industry was about to enter a new era, where its relationship to living plants grew ever more tenuous.

Édouard Laugier would not be a part of it. While he continued to live in the neighborhood where he grew up, with a shop on rue Greneta, successive descriptions in the commercial almanac show how far he had moved from the family business: from distilling, to bottling, to wine corks, to cork shoe soles. Unmarried, with no children, he died on 14 March 1869, leaving precious few clues for future historians to tease out the details of his later life.[70]

12

Life Is Asymmetric

LONDON, 1855

August Hofmann was more determined than ever to break into the market of synthetic scents. He had been the first to envision a new era of perfumes made from coal tar, and he wanted to make sure he did not miss out on any fortune to be made. He tightened his focus on compounds related to benzoic acid and coal tar, as well as essential oils he wanted to replicate. He began to notice certain characteristics common to all these compounds: they had, relative to other organic molecules, much more carbon and less hydrogen. And the carbons always seemed to come in groups of six, never going below this minimum number. In 1855, he named this class of compounds "aromatics," convinced they were the key to the chemistry of scent.[1]

There was going to be another Universal Exposition that year, this time in Paris. The British hoped that they could regain the ground they had lost to Collas in 1851. Hofmann's student, Charles Blachford Mansfield, was particularly determined to have an impressive showing still smarting as he was from Collas's use of his patented processes. Mansfield had set up his own business by now, working out of a detached building on Regents Canal. He supplied Hofmann and others with nitrobenzene and aniline, although by February of 1855 he was concentrating on making perfume samples for the Paris Exposition. On the afternoon of 17 February, around 1 p.m., he was distilling naphtha with his eighteen-year-old assistant, George Coppin, when the still exploded, destroying the building and engulfing them in flames. Onlookers saw the two men running from the wreckage, clothes on fire, toward the canal, which was then frozen over. They rolled themselves onto the ice, and eventually, with help, extinguished the flames. Both were taken to the hospital, where they held on for a few days more, although they were already so badly burned that they resembled shriveled mummies. When Hofmann came to visit, Mansfield greeted him with the phrase, "here lie the ashes of Charles B. Mansfield."[2] He died soon after, at the age of thirty-five.

Chemistry was no discipline for the faint of heart. But somehow there were always more than enough hopeful students willing to take the risk. Hofmann still needed aniline, so he assigned the task to his youngest assistant, William Henry Perkin, who had joined the lab two years earlier, at the age of fifteen. A bright and curious boy, Perkin had always been more inclined toward art and music, but when he was thirteen he saw a friend perform some experiments with crystals and was hooked.[3] Much of his work at the Royal College was in support of Hofmann, but when Hofmann left for a visit to Germany during the Easter holidays in 1856, he began a project of his own: trying to make artificial quinine from aniline. The appeal was obvious. Demand for quinine had never been higher. It was the crucial ingredient for the medicinal gin and tonic taken nightly around the British Empire. It had been one of the first goals of Laurent and Hofmann in their earliest work together. But Hofmann had already tried, with Liebig, and the result was an economic debacle.[4] With the optimism of youth, Perkin tried his own manipulations of aniline, adding some oxygen and removing some hydrogen. But where he hoped to see the colorless quinine appearing in his flask, he got instead something thick and tarry black.

And yet Perkin noticed something interesting in his failed experiment. Purified and dissolved in wine spirits, the blackish substance took on a pretty violet hue, and when Perkin stained a cloth with it, the color remained strong and vibrant even after he washed it and exposed it to light. Before Hofmann even returned from his trip, Perkin had fitted out a hut in his garden to begin making larger batches. He sent off samples to a dye works in Scotland, and after a positive response, took out a patent on the process. Dyes at the time came mostly from plants and occasionally an animal source like the cochineal insect. Purple was particularly difficult to achieve—both very expensive and susceptible to fading. There was some concern that at eighteen, Perkin's age might be a problem, but the patent went through. He called it Tyrian Purple.

By the time Hofmann came back to England, Perkin had decided to leave the Royal College of Chemistry to make his dye on an industrial scale. It took 100 pounds of coal to make a quarter-ounce of dye, as he ushered it through the steps of transformation and distillation, from nitrobenzole, to aniline, to marketable product. As he began production, his streak of good luck continued. Over in France, where style was set, Napoleon III's wife, the Empress Eugénie, had decided in 1857 that her signature color was going to be a delicate purple that Lyon silk manufacturers Guinon, Marnas et Bonnet had recently produced from ground lichen. They named it "mauve," and it soon became the most sought-after color, nearly impossible to get because of its scarcity. Perkin, seeing an opportunity, renamed his product "mauveine" and offered it at a fraction of the price of the natural product. When Queen Victoria wore a mauve dress to the wedding of her daughter in 1858, demand reached a fever pitch.

Hofmann had initially been suspicious of what he called Perkin's "purple sludge." But he was not one to ignore success, and soon produced his own dye, "Hofmann's violets." The following years saw a cavalcade of new, synthetic colors: Verguin's fuschine, Manchester brown, Bismarck brown, Martius yellow, Magdala red, Nicholson's blue. It was, Hofmann said, "the strangest of revolutions." Where England used to spend millions importing exotic ingredients from across the globe, the new aniline dyes meant that it was the one now sending "her coal-derived blues to indigo-growing India, her tar-distilled crimson to cochineal-producing Mexico, and her fossil substitutes for quercitron and safflower to China and Japan."[5]

The age of synthetic chemicals had begun, kicking off what came to be known as the second industrial revolution. The first industrial revolution had centered on the steam engine and its remarkable ability to use the energy stored in fossil fuels to replace human labor. And now the second hinged on a similar ability to exploit fossil fuels, only this time they replaced the labor of the many plants (and a few animals) that worked to turn energy from the sun into useful organic compounds. It was a marvelous shortcut that promised to make rare things available to all. Fossil expert (and inventor of the term "dinosaur") Richard Owen saw the situation clearly. "Already, natural processes can be more economically replaced by artificial ones in the formation of a few organic compounds . . . It is impossible to foresee the extent to which chemistry may ultimately, in the production of things needful, supersede the present vital energies of nature."[6]

Hofmann had expected perfumes to be the most important products of his aromatic compounds, and had been surprised at the quick success of aniline dyes. But artificial perfumes did follow in their turn. Perkin was once again a central figure. And, once again, he got there by accident. As he was trying to replicate an experiment with salicylic acid that the French chemist Cahours had described in 1857, he swapped out the hydride of salicyl with its sodium derivative. He was no closer to his intended product, but noticed that whatever he had made smelled really good. He recognized the sweet, slightly haylike smell immediately: tonka beans, a long-standing perfume ingredient. Wanting to be sure, Perkin purchased some tonka beans and extracted the substance coumarin, responsible for its distinctive smell. He compared it to his new creation. Their odor, he indicated, was "exactly the same," and he began to call his substance "coumarine."[7] He had done it again. The *New York Herald* dubbed him the "Coal Tar Wizard" who had "Transmuted Liquid Dross to Gold."[8]

By this time, Hofmann had moved back to Germany, accepting a position at the University of Berlin. There, too, he was surrounded by young assistants, including Wilhelm Haarmann and Ferdinand Tiemann (whose sister, Bertha, would soon become Hofmann's fourth wife). In Hofmann's lab, they were able to synthesize a new molecule from pine bark that smelled just like vanilla. Unlike coumarin, it did not seem to resemble any of the compounds actually found in vanilla beans, but the smell was so similar they were sure it could serve as a satisfying substitute. They named it vanillin, and within the year, Haarmann had opened a factory producing it on a mass scale.

A similar story had played out a hundred or so miles west, in Göttingen, where Rudolph Fittig worked as an assistant to Friedrich Wöhler and had recently completed his thesis on the contributions of Auguste Laurent to organic chemistry (in death as in life, Laurent remained better appreciated outside his own country than within it). In 1869, he teamed up with W. H. Mielk to synthesize a molecule they named piperonal.[9] It had a soft, powdery scent, reminiscent of the dainty purple flower heliotrope. Heliotrope itself was not used much in perfume, and it turned out there was no piperonal in heliotrope, but the opportunity was too good to pass up. They renamed it heliotropin, and it soon began to appear in soaps and perfumes.

In France, it was the chemist Georges de Laire who first moved into the world of synthetic perfumes. Trained under Pelouze, he had begun working with dyes while at Hofmann's lab in London.[10] He had gotten rich off the French patents for synthetic dyes, but after working with Haarmaan and Tiemann on vanillin, he switched to the production of perfumes.[11] He used all the new synthetics available, and even developed an artificial musk himself. As industrial production ramped up, the price of synthetics plummeted. From 1879 to 1899, the price of a kilogram of heliotropine dropped from 3,790 to 37.5 francs. Coumarine fell from 2,550 francs for a kilogram in 1877 to 55 francs in 1900, and a kilogram of vanillin dropped from 8,750 francs in 1876 to 60 francs in 1906.[12]

The real turn of the tide occurred in 1880, when Houbigant, one of oldest and most powerful perfume houses, began to experiment with synthetics. Its new owner, Paul Parquet, took Perkin's coumarine as his base, with its lingering scent of fresh-cut hay, and layered the sweeter notes of geranium, rose, and lilac as its heart, and lighter notes of lavender, bergamot, and chamomile as its head.[13] The result had a mossy, lush greenness that prompted him to name it Fougère Royale, or royal fern. He chose the name precisely because royal ferns themselves had no scent, making it a perfect template for an artificial creation. ("If God gave ferns a scent, they would smell like *Fougère Royal*," he was famously noted as saying.)[14]

The new perfume was an immense success, particularly among men, who seemed to find it more acceptable to smell like ferns than flowers. It went on to inspire a whole family of *fougères* alongside the florals and woods, all organized around the evocation of a scent that did not exist. Soon, nearly every house was using synthetics, which were seen not as cheap replacements, but as a welcome expansion of the perfumer's palette. The fantasy scents they created, increasingly distant from any particular object of natural origin, were the next frontier of luxury objects.[15]

In London, the satirical magazine *Punch* replaced the lyrics of the well-known song "Beautiful Star" with one they called "Beautiful Tar—song of an enthusiastic scientist."

> *There's hardly a thing that a man can name*
> *Of use or beauty in life's small game*
> *But you can extract in alembic or jar*
> *From the "physical basis" of black Coal-tar.*[16]

"Triumph O Tar! Stuff half divine!" sang the chorus. But not everything made from coal was a success. Mirbane, the first artificial perfume, had turned out to be a disaster. The nitrobenzol that smelled so much like bitter almonds was, as it happened, incredibly toxic, as was the benzene that Collas peddled as a stain remover. Doctors noted in tests that it killed rabbits and dogs treated with it.[17] People who used it regularly noticed that their formerly rosy cheeks took on a "dirty leaden color" that eventually became black or dullish blue around the lips, tongue, and ears.[18] Their urine turned dark and smoky, looking like port wine and smelling like bitter almonds. Their blood darkened to nearly black, and their hands and feet tingled. They complained of headaches, soreness, and an unshakable tiredness. It was, doctors said, some kind of anemia caused by the chemicals attacking the bone marrow. They counted dozens of cases of poisoning, at least thirteen of them fatal.[19] Sales of mirbane and household benzene dwindled, but the question remained: were these artificial creations the same as their natural counterparts? Were they safe?

PARIS, RUE SAINT-JACQUES, 1860

By 1860, Pasteur had been leading a secret life for over ten years, and he was ready to come clean. By all outward appearances, he was one of the most accomplished scientists of his generation. After his work with Biot, he had taken teaching positions first at the University of Strasbourg, then Lille. By 1857, he had returned triumphantly to Paris as the director of Scientific Studies for the École Normale. His crowning achievement had been figuring out the process of fermentation, succeeding precisely where Lavoisier had failed. Lavoisier had been convinced that fermentation was a fundamentally chemical process, and despite his inability to prove it, most of his followers agreed, mocking vitalists like Chaptal as naive.[20] Yet in the end, it was a "vital act," as Pasteur put it.[21] The ferments in the yeast turned out to be living things—"microbes" or "germs."[22] Pasteur showed the necessary role they played in the process, and the Academy of Sciences honored his work with the 1859 Prize for Experimental Physiology.

But privately, he struggled with a task that continued to elude him. His work on fermentation, he hinted to his closest friend, Charles Chappuis, was itself a path to "the impenetrable mystery of Life and Death."[23] He was on the trail of some mysteries, he confided, and "the veil that covers them is getting thinner and thinner."[24] It all hearkened back to the connection he had learned of from Laurent between a compound's crystalline form and its optical activity. He rarely spoke Laurent's name anymore, and it appeared to outside observers that he had given up any of the work they shared. But he had simply gone dark about it, working secretly in his lab. It had, in fact, been the optical activity of tartrates, those crystals produced by fermenting wine, that led him to study fermentation in the first place, and convinced him that the chemical approach was wrong. Liebig had claimed that fermentation was a form of disintegration, with the sugar broken apart by decay. He saw tartrates as leftover debris, which preserved its activity from the original sugar. But Pasteur's experience was that optical activity disappeared at the slightest change in molecular structure and that life alone could produce such asymmetries. When his work on fermentation confirmed his suspicions, he wanted to see if this was part of a more general pattern.

Pasteur probed the boundaries of the natural and synthetic, spending hours doing secret experiments to see if he could ever make something, *anything*, in his lab that was optically active.[25] He found that he could not. The closest he ever came was when he tried to force the issue and "introduce asymmetry into the chemical actions of the laboratory."[26] He combined cinchonicine (an active substance) with paratartaric acid (an inactive one). He was able to separate out the left tartrate, leaving the right tartrate behind. He had thus "made" an active substance from an inactive one, but as it required him to start with the active cinchonicine, he was left more convinced than ever of the "barrier" between natural and artificial chemistry, and that "the forces in play in our laboratories differ from those which govern vegetable nature."[27] But these results remained unpublished, recorded only in Pasteur's notebooks.[28]

By 1860, he was ready to go public, and he chose the occasion of the inaugural lecture of the Chemical Society of Paris. This was a new group, recently formed in the model of the Chemical Society of London, and they put together a well-publicized lecture series open to the public. Dumas was its president, and he invited Pasteur to give the first lecture on January 20. Pasteur, obliging, provided the title, "On the Asymmetry of Naturally Occurring Compounds," and prepared to reveal the secret he had been keeping for years: the work on crystals and optical activity revealed a deep, unbridgeable gap between living and inert substances.[29]

Pasteur was nervous about the presentation. He had to make a delicate, complex argument that required knowledge of not only chemistry but physics and crystallography as well. So he began with a small primer on these subjects, going all the way back to the discovery of the polarization of light in 1809 and detailing Biot's work with essential oils to establish that organic matter acted on light at the molecular level. He next taught the audience the basics of mineralogy, explaining how Haüy had identified certain crystals as having an asymmetric, "hemihedral" arrangement. Like crystals, molecules had their own internal arrangement, with their own symmetries or asymmetries. Some molecules were identical to their mirror images, the way something symmetrical like a square would be. Some molecules, however, were fundamentally different and nonsuperimposable to their mirror images, like a pair of human hands. Pasteur recounted how he had found this sort of hemihedral asymmetry in tartrates, including the anecdote in which Biot took his arm and spoke of his beating heart. Pasteur even provided a dramatic demonstration at the front of the lecture hall, where he mixed together some tartrates that deviated the plane of polarization to the right with some that deviated the plane to the left, and combined them to form an inert, artificial paratartrate, which did nothing to the light at all. The audience erupted in applause. He had not even gotten to the best part yet, but he was out of time and would have to save the remainder for a second lecture, scheduled two weeks later.

They reconvened in the first week of February, and Pasteur gave away the grand reveal. It was remarkable, he said, that every artificial molecule made in a laboratory, or found in the mineral realm, was identical to its mirror reflection, while on the contrary almost every natural organic molecule was not. ("I could even say all," added Pasteur, "if I were only talking about those that play an essential role in the phenomena of vegetable and animal life."[30] Chemistry, he said, should be divided into distinct realms of "living" and "dead" substances.[31] The standard division between organic compounds, which contained carbon, and inorganic, which did not, was insufficient and misleading. It made no distinction between organic molecules that occurred naturally and those created in a lab. And there was a difference between them, present in the asymmetry of the molecular arrangement. Polarized light could alone reveal this asymmetry, and for Pasteur, optical activity became the arbiter of the division. Substances that were optically active were "living nature"; substances that were not were "dead nature." True, there were some naturally occurring organic substances, such as benzene and urea, which were optically inert, but these, Pasteur said, should be considered "excretions rather than secretions." They were not part of the vital "organic juices" that played a role in organization (what he called "the immediate principles essential to life") but were more of a waste product, discarded in passing.[32]

He voiced the question on everyone's mind: "Why this asymmetry?" Pasteur could not say. But asymmetries did happen in nature. Physicists were perhaps more familiar with them than chemists, having already come to terms with the fact that the planets revolving around the sun all went in one direction and not the other, or that, according to Ampère's Law, a current passing through a wire always created a magnetic field in a right-hand rotation around it, and never a left-hand one. Just as some people thought that the rotation of the planets offered a clue about the origins of the solar system, Pasteur also thought his findings pointed to some asymmetric force present at the moment of what he called "the elaboration of the immediate principles in the vegetable organism," or the origin of life. "Can these asymmetric actions be connected with cosmic influences?" he asked.

One thing was clear to Pasteur: this action of nature, whatever it was, could not be reproduced in a lab. While the fall of sunlight upon green leaves gave birth to dissymmetric compositions, even the most skilled chemist, using all the techniques of synthesis at their disposal, could only make symmetrical products. "No. Chemistry has never made an active compound from inactive products. . . . Chemistry will remain powerless to make sugar, quinine, [and other immediate principles of life] so long as it continues on the erroneous path of its current procedures."[33]

Pasteur was pleased with how the talk went, writing to his father that he presented his ideas with "clearness and power, and everyone was struck by their importance."[34] Dumas, who chaired the event, rose afterward to praise him for the enthusiasm and novel ideas.[35] But he remained wary of the new direction Pasteur was taking. Pasteur had seen his efforts as bringing "a little stone" to "the frail and ill-assured edifice of our knowledge of those deep mysteries of Life and Death where all our intellects have so lamentably failed."[36] He wanted to go further down the path, but Dumas warned him away—"I would advise no one to dwell too long on such a subject," he said.[37] Even the newspapers doubted him, as *La Presse* wrote, "The world into which you wish to take us is really too fantastic."[38]

Biot was not at the talk, but he heard about it the next day and let Pasteur know he was delighted. A seat opened up at the Academy of Sciences in the Botanical Section, and Biot lobbied hard for Pasteur's election. The botanists were skeptical. One proposed that they go to Pasteur's house, and if they found a single book on botany in his library, he would support him.[39] He did not get the position. But there was an opening the next year in the mineralogical section, and they voted him in. In 1863, Dumas brought him to the Tuileries and presented him to the emperor, who received him rather distractedly while Pasteur pledged his intent to fight against putrefaction and disease.

Soon he was commanding the kind of crowds he had only seen before with Dumas. The Sorbonne chose him as their first speaker to launch a series of scientific and literary "soirées" that drew the fashionable elites of Paris. The audience that filled the enormous auditorium included George Sand, Princess Mathilde (the emperor's cousin and noted salonnière), and the minister of education, Victor Duruy. Alexandre Dumas was there, too. Already famous for his novels *The Count of Monte Cristo* and *The Three Musketeers,* the author had long joked about being the secondmost respected Dumas in Paris. The chemist, he pointed out, was so widely known as "Dumas the savant" that he took to referring to himself as "Dumas the ignorant."

Pasteur asked a single question in his talk: "Can matter organize itself?"[40] His answer was an emphatic no. He took particular aim at the theory of spontaneous generation, the age-old belief that living creatures could arise from inanimate matter. It had seen a recent revival in conjunction with the ideas of evolution then becoming fashionable, particularly those of Lamarck, who thought that life was continuously emerging from inert matter, simple at first but growing more complex with each generation. Proponents of this theory liked to point to the following observation: if you let a cask of water sit for a day or two, then return to look at a sample through a microscope, you would see a whole world of writhing "animalcula" or "living atoms," no matter how rigorously you kept any foreign creatures from laying their eggs inside.[41] Pasteur demolished this claim in his talk. He demonstrated with his specially made "swan-necked" flasks that as long as you kept any dust out of the flask, no microscopic beings appeared. The dust, he proposed, carried with it "organized corpuscules" or "infusory animalculae," which served as the origin of any life that appeared. There was no chance of getting life from nonlife. The inviolable chasm remained intact.

Pasteur was only forty-two, and his most spectacular accomplishments still lay ahead of him. Over the next decades, he would save the French silkworm industry from being wiped out by disease, develop vaccines for rabies, anthrax, and cholera, propose the germ theory of disease, and invent the process of pasteurization to preserve milk and wine. His contributions to the public good have been so staggering that he has achieved a status akin to sainthood in the pantheon of French scientists. But however varied his later endeavors were, they retained the kernel that he had kept secret from his fellow chemists for so long: life was a mysterious thing with its own inimitable organization, not reducible to a chemical formula alone.

THE LEGACY

It may have seemed like madness for Biot to devote himself to finishing the work of someone like Laurent, who had been so studiously ignored in his lifetime. But Biot approached the task with reverence, acknowledging in the preface that these were "intimate convictions" that Laurent had labored on "until in the arms of death."[42] The end result was the book *Method of Chemistry*, Laurent's strange, bold reimagining of chemical classification. It sought to save chemistry from its perverse flaw—that it studied only objects that had ceased to exist and were already broken into their constituent pieces. Instead, Laurent focused on arrangement or organization, grouping compounds in series based on their molecular form. He had been just about the only chemist, Biot pointed out in his preface, to use polarized light as a tool for exploring a molecule's shape. The book dealt with 7,000 to 8,000 compounds, all reclassified and given new names. In his own idiosyncratic twist, Laurent included drawings of certain molecules, including, as a variety of the essence of bitter almonds, the "complete polyhedron" of benzoyl chloride.[43] It was an unprecedented thing to do, depicting this preparation of benzene as a hexagon, and most people ignored the entire enterprise.

One person who read it was August Kekulé, in preparation for writing a textbook of his own. An aristocratic, cosmopolitan German who spoke four languages, he was less doctrinaire than many of Laurent's critics. "I no longer belong to any school," he claimed, after a particularly peripatetic education. First "seduced" into chemistry by Liebig, he spent four years in Giessen, then left for a grand tour of the chemistry labs of Europe.[44] He spent 1851–1852 in Paris, attending Dumas's lectures but working most closely with Gerhardt.[45] He spent 1853–1855 in London and eventually settled in Ghent, Belgium, where he took a position as a professor of chemistry.

The story, as Kekulé told it, was that one evening in 1862, after a long day working on his textbook, he retired to his bachelor apartment and settled into a warm chair before the fireplace. He was half dozing as he began to see atoms flutter before his eyes, connecting themselves into long chains which, twisting and turning, appeared to be snakes. "But look, what was that? One of the snakes seized its own tail, and the figure whirled mockingly before my eyes."[46] Kekulé reported that he awoke as if struck by lightning and worked out the consequences of his dream: the atoms of benzene looped back on themselves, cyclically connected like a snake biting its tail. Benzene, he realized, could only be understood by its structural arrangement, a six-sided hexagon.

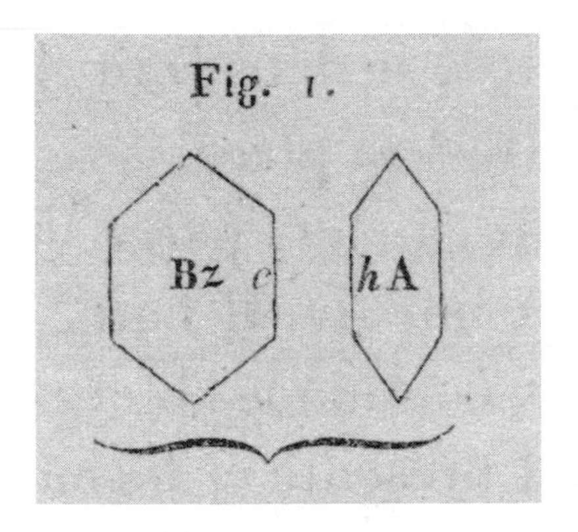

FIGURE 33. Auguste Laurent's drawing of the "complete polyhedron" of benzoyl chloride in his textbook, *Méthode de chimie.*

This moment, immortalized as "Kekulé's dream," has become one of the most iconic in the history of chemistry. Kekulé only told the story once, however, nearly thirty years after the fact at a celebration called Benzolfest held in his honor in 1890. There has been some suspicion that Kekulé, by saying it came to him in a dream, was able to sidestep the question of any intellectual influences, particularly if they came from so despised a source as Laurent.[47] In any case, the very idea that had garnered Laurent so much ridicule—that molecules should be thought of as a spatial arrangement of atoms—now exploded like a thunderclap of profound insight. Kekulé became the master of a new kind of structural chemistry. He devoted his lab to studying these arrangements, using Tinkertoy-type models, with colored wooden balls representing different atoms and thin brass rods as the bonds connecting them.[48]

By 1872, chemists from across Europe had come to work with Kekulé in his laboratory. The youngest among them was a Dutch student named Jacobus van 't Hoff, who had sought out Kekulé as the leading authority of structural theory. While Tinkertoying around and following the rules that he thought atoms should follow, he noticed something remarkable: in certain cases, when a carbon had different clusters of atoms attached to its four corners, he could make two distinct models representing the same chemical formula. They were perfect mirror images of each other, but not superimposable, much like the right- and left-handedness Pasteur had pointed to. It was a strange sort of asymmetry, and he suspected it was responsible for the optical activity that rotated light in different directions. He rushed to check against a long list of optically active compounds. Sure enough, they all had one of the asymmetric carbons in question.

Barely twenty-two, van 't Hoff self-published a pamphlet called *Chemistry in Space*.[49] It was an argument for the importance of the spatial arrangement of atoms, and not, as it sounds, the latest Jules Verne novel. But it may as well have been, for the reaction it provoked. Van 't Hoff had by then graduated and, for lack of better prospects, taken a job at a veterinary school. One of Germany's top chemists, Hermann Kolbe, best known for his fight against vitalism, snidely suggested that van 't Hoff, who had "no taste for accurate chemical research," had instead borrowed a Pegasus from the veterinary stables to ride to the top of Mount Parnassus and boldly proclaim "how the atoms appeared to him to be arranged in cosmic space."[50]

France was equally hostile to the Tinkertoy approach. Dumas was still the most powerful influence at the Academy, and he continued to discourage the approach of treating atoms as real things that could be modeled in space. But van 't Hoff would soon get support for his ideas from an unexpected source. A young French chemist, Joseph Achille Le Bel, had been working as an assistant to Balard but could not put Pasteur's asymmetries out of his mind. As he continued to think about what kind of molecule could produce the results Pasteur described, he eventually zeroed in on the same structure van 't Hoff had identified: a central carbon atom with different atomic groupings attached to it. He had come to the idea independently and published it only weeks after van 't Hoff had put out his pamphlet. It was as if the two components of Laurent's thoughts—the insistence on spatial arrangement and the focus on optical activity—had come crashing together to finally explain what was going on: the atomic clusters attached to the carbon formed wayward "hands" that came in left and right mirror images. Chemists eventually acknowledged the accuracy of their claims and devoted a new field, "stereochemistry," to its study. Van 't Hoff himself eventually left the veterinary school for a better job and won the first Nobel Prize ever awarded in chemistry.

It would have been a tremendous vindication for Laurent if he had still been alive (and he would have been only sixty-seven). It seemed to tie everything together nicely, with a bow. But a deep, nagging question still remained: why did living things always seem to come in a one-handed variety alone, while artificial things produced both kinds of handedness indiscriminately? Nobody knew. Pasteur's dictum that "life is asymmetric" remained true, and his bright line between life and death remained intact. Research pressed on.

Britain's most eminent physicist, Lord Kelvin, coined the term *chirality* to describe the situation, borrowing from the Greek word for "hand." The occasion was an 1884 lecture series in Baltimore that attempted to lay out the basics of the new electromagnetic theory of light and its interactions with matter. He pointed out that there was still much work to be done, and two of his audience members, Albert Michelson and Edward Morley, were inspired to build an interferometer to look for the ether he kept talking about light travelling through.[51] Kelvin also hoped to inspire chemists to take up the question of chirality. He explained its physics: the handedness of molecules set up electric fields that would turn the plane of polarization either to the right or to the left. But chemists faced a substantial problem. The different chiral variants were, from a chemical point of view, identical, with the same weight, density, solubility, melting point, boiling point, viscosity, or any other characteristic chemists could measure. In chemical reactions with symmetric, achiral molecules, there was no perceptible difference between them.

There was one extraordinary, well-tuned device that could distinguish immediately between two mirror-image molecules: the human body. Like all living things, our bodies are composed of chiral molecules that all came in a single variety. And as such, they respond completely differently to right- or left-handed versions. (Imagine trying to shake hands—if you are offering your right hand, it matters whether you are getting a right or left hand in return.) Bodies can only metabolize, for example, sugar of the right-handed variety. And while the right-handed version of some substance may do something useful like reduce a fever, the left-handed version might do the opposite or worse. This became a big issue once the artificial production of drugs opened the pandora's box of molecules mirror-imaged to their natural counterparts. In some cases, everyone got lucky. The first widespread artificial drug, acetyl salicylic acid, turned out to be completely symmetric, with no chiral centers distinguishing one mirror-image from the next. Laurent's old partner, Gerhardt, had been the first to synthesize it in 1853, from willow bark, but it became much more widespread when Bayer began mass producing it under the name "aspirin."[52] The later drug ibuprofen had similar luck. It was chiral, and only the left-handed versions were responsible for its salutary effects. The right-handed versions were fortunately harmless, passing through the body without doing much.

But chirality could count in devastating ways, as the example of Thalidomide revealed. It was a wonder drug of the 1950s, prescribed particularly to pregnant women to control their morning sickness. Yet while the left-handed version was safe to take, the right-handed version was teratogenic, disturbing fetal growth. The two versions were mixed indiscriminately in the lab-created medicine, ultimately affecting some 10,000 babies. Only half survived, and those that did had substantial birth defects, most notably marked by stunted arms and legs. Amid the scandal, chemists had a new appreciation of how these mirror-images, which they now called *enantiomers,* could affect the body differently.

By far the most extensive way that the human body interacts with the chemicals around it is through the sense of smell. The nose sifts through thousands of chemicals a day, a process so taken for granted that it is hardly even thought about. It seems to have a virtually infinite ability to distinguish between molecules, including, remarkably, between two enantiomers, which can seem to the nose to be completely different scents. The molecule carvone is a prime example, as it is somehow responsible for both the spicy pungency of caraway seeds and the cool sweetness of spearmint. In one of the cases of scientific serendipity that are more common than one thinks, three different research groups cracked the case of carvone simultaneously. They worked out the spatial arrangement of the constituent atoms and showed that the two versions were flipped mirror-images of each other, establishing conclusively that enantiomers smelled different.[53]

As perfumers moved deeply into the world of synthetic fragrances, they navigated the tricky world of enantiomers and pioneered the complex science of fragrance chemistry.[54] Their research invited the question of how, precisely, the nose is able to recognize hundreds of thousands of different molecules. Does it recognize vibrational frequencies? Do the molecules "dock" to certain receptors that fit their precise shape? Enantiomers seem to offer a crucial clue that shape plays a role, but the question is far from settled.[55] Smell remains to this day the least understood of all the senses. Scientists have only recently begun to realize how complex it is—far more complex than vision, and more intimately connected to the brain.[56] This renewed interest in smell reverses Enlightenment tradition, which generally associated vision with a person's rational faculties and smell with their primitive, animal-like impulses.

But if Enlightenment thinkers had dismissed the sense of smell, alchemists had not. The *spiritus rector* had been a vital component of living things before Lavoisier dismissed it as a fictional category. It had functioned as a "guiding spirit" that gave shape and organization to an organism, elevating it beyond mere matter. The chemical revolution had sought to strip it of its power and to explain the processes of life as straightforward reactions of precisely the same chemicals that made up the inorganic world. But amid its numerous successes were some failures, too. And most persistent among these was the puzzling, unexplained asymmetry of life. Indeed, it seemed as if some guiding spirit retained a role in its organization.

Could life organize itself? Pasteur had said no. The subsequent discovery of DNA in the 1950s revealed the molecular basis for how the organizational blueprints get transmitted. But there still remained the question of how this arose in the first place. And there was one more puzzle as well: the sugars that made up the DNA backbone were all chiral in one direction only. And the amino acids it encoded were as well, with every one of them being left-handed (with the sole exception of the achiral amino acid glycine). Indeed, the presence of any right-handed ones would have prevented the system from working. Scientists coined the term *homochirality* to describe the situation, but they were no closer to explaining it.

Pasteur, in his revelatory speech of 1860, had already speculated that "cosmic forces" might account for it. And in the ensuing 150 years, astrophysicists have uncovered a number of intriguing asymmetries in the cosmic environment. The spiral motion of the Milky Way, for example, changed the circular polarization of the light moving through it, which in turn could pass this handedness on to amino acids. The weak nuclear force was also known to be asymmetric, prompting the decay of electrons whose spin was always only left-handed. Perhaps these charged particles, raining down on earth as cosmic rays, affected life.[57]

But biochemists are skeptical that such subtle mechanisms could account for such a stark divide. After all, life on earth showed not just a preponderance of one chiral type, but an utter, complete monopoly, without a single mirror-image copy surviving. Perhaps, some proposed, the appearance of chirality actually preceded the origin of life, and was in fact a necessary antecedent to it. The original primordial soup on Earth likely had amino acids of both chiralities scattered about at random. But it would have been impossible to get life this way. The next step, most people agree, would have been for them to come together to form larger chains of RNA, a process known as polymerization. But the presence of different chiral versions sabotaged this process because of something called enantiomeric cross-inhibition, which kept long chains from forming. The only way for life to begin was to break this symmetry. There had to be a moment of "breaking the mirror" that let one mirror-image win out over the other before the evolution of life could proceed.[58]

Perhaps it was nothing more than chance that in a tiny pond somewhere a few billion years ago, an abundance of left-handed amino acids got the process started. In that case, was it possible that on some other planet, if things went just a little bit differently, there might be a whole "Mirror World" of organisms with right-handed amino acids and everything else an inverted reflection of our own world? To settle the question, scientists looked for any form of extraterrestrial life available to them.[59] The Murchison meteorite fell in Australia in 1969. It was enormous, and riddled with amino acids and other "prebiotic" molecules. Were they homochiral? It was hard to say. At first, the amino acids seemed to be evenly distributed, left and right. Later samples were found that did have more of one enantiomer than the other, but there was some concern they were contaminated by earthly bacteria. The only way to be completely sure there was no contamination was to get a sample that had never been on Earth—not the easiest proposition. But when the European Space Agency landed the probe Philae on a comet in 2014, scientists hoped to finally get an answer. They had been able to determine that the comet had several prebiotic molecules, including at least one amino acid, using spectroscopic data, acquired remotely. To test the chirality, however, they required a sample of actual material from the comet. Unfortunately, the probe bounced as it landed and ended on its side, wedged atop its now inoperable drill. It is still there, the comet now millions of miles away, and the question of life's asymmetry is left unanswered. "We may never know," admits Laurence Barron, one of the leading researchers.[60] There is mystery to life, yet.

Cast of Characters

ANTOINE BAUMÉ (1728–1804) A French Enlightenment chemist who gave public demonstrations of chemistry and distilling with Pierre Macquer.

JEAN-BAPTISTE BIOT (1774–1862) A cantankerous, iconoclastic French physicist who studied the polarization of light.

ALEXANDRE BRONGNIART (1770–1847) The wealthy and politically influential director of the Sèvres Porcelain Factory. His daughter Herminie married Jean-Baptiste Dumas.

ÉLISABETH-FÉLICIE CANARD (1796–1865) An independent, bookish woman who wrote manuals on style and grooming under the pen name of MADAME CELNART.

JEAN-ANTOINE CHAPTAL (1756–1832) A chemist from the south of France who was particularly committed to the idea that nature is animated by vital forces.

CLAUDE COLLAS (1810–1876) A business-minded French pharmacist who pioneered the use of "essence of mirbane" as an artificial substitute for bitter almond.

CATHERINE DUFRAYER A perfumer, she married Jean Laugier in 1796 and continued to work in Paris after his death.

JEAN-BAPTISTE DUMAS (1800–1884) A charming and savvy French chemist, director of the École Centrale des Arts et Manufactures, who turned his outsized ambitions to conquering the organic realm.

JEAN-MARIE FARINA (1685–1766) The original purveyor of Eau de Cologne, with a store across from Jülichs Platz in Cologne, Germany.

JEAN-MARIE-JOSEPH FARINA (1755–1864) A distant, perhaps false relation of Jean-Marie Farina who arrived in Paris in 1808 and copyrighted the name "Farina" and other terms associated with Eau de Cologne.

ANTOINE-FRANÇOIS FOURCROY (1755–1809) Lavoisier's closest collaborator, whose flair for public speaking made him the "apostle of the new chemistry."

CHARLES GERHARDT (1816–1856) A firebrand chemist from Alsace who worked in the labs of Justus Liebig and Jean-Baptiste Dumas before persuading Auguste Laurent to join forces with him.

AUGUST HOFMANN (1818–1892) Justus Liebig's assistant at Giessen who was handed an outsized opportunity as the inaugural director of the Royal College of Chemistry in London.

ANDRÉ LAUGIER (1770–1832) A pharmacist who worked with Antoine-François Fourcroy; no relation to the perfumers on rue Bourg-l'Abbé.

BLAISE LAUGIER (1737–1826) A perfumer from Grasse who set up shop on rue Bourg-l'Abbé in Paris.

ÉDOUARD LAUGIER (b. 1807) The son of Jean Laugier and Catherine Dufrayer who left the family's perfume business to become a chemist.

JEAN LAUGIER (1768–1836) Blaise Laugier's oldest son, who ran the Paris shop when he retired.

MARIE-JEANNE (FEURER) LAUGIER (d. 1800) Blaise Laugier's wife. She had seven children: Jean, Louis, Madeleine, Antoine François, Alexis, Blaise Jr., and Auguste Victor.

AUGUSTE LAURENT (1807–1853) An artistic wine merchant's son who left the career of a mining engineer to study chemistry, a theoretician and revolutionary who was frustrated in his ambitions by the distrust and animosity of Justus Liebig and Jean-Baptiste Dumas.

ANTOINE LAVOISIER (1743–1794) A fabulously wealthy tax collector who led the chemical revolution, overturning centuries of knowledge.

JUSTUS LIEBIG (1803–1873) A brilliant but highly volatile chemist who ran a laboratory of practical chemistry in Giessen.

PIERRE MACQUER (1718–1784) An Enlightenment chemist who gave the theoretical component of the public lectures with Baumé, laying out a theory of aroma as the *esprit recteur*, or guiding spirit.

CHARLES BLACHFORD MANSFIELD (1819–1855) August Hofmann's student, who hoped to make a fortune in the perfume business.

LOUIS PASTEUR (1822–1895) A student working in a chemistry lab at the École Normale who grew curious about Auguste Laurent's strange experiments with crystals. He went on to be one of the most prominent scientists of France, credited with saving millions of lives.

WILLIAM HENRY PERKIN (1838–1907) An entrepreneurial teenaged assistant of August Hofmann at the Royal College of Chemistry in London who developed mauveine, the first commercial synthetic dye.

PIERRE-JEAN ROBIQUET (1780–1840) An assistant to Nicolas-Louis Vauquelin who grew obsessed with the relation between cyanide's deadly effects and its pleasant aroma.

EUGÈNE ROUSSEL (1810–1878) Successor of Édouard Laugier and Auguste Laurent at Laugier Père et Fils who later emigrated to Philadelphia, where he became wildly successful in the new flavored soft drink business.

LOUIS JACQUES THÉNARD (1777–1857) A student of Antoine-François Fourcroy and Nicolas-Louis Vauquelin who went on to hold nearly every available chair of chemistry in Paris.

NICOLAS-LOUIS VAUQUELIN (1763–1829) An assistant of Antoine-François Fourcroy who helped him define the field of pharmacy, and the mentor of Louis Jacques Thénard.

FRIEDRICH WÖHLER (1800–1882) Liebig's frequent collaborator, who sometimes encouraged and sometimes tempered his biting sarcasm.

NOTES

PROLOGUE

1. J. Liebig and F. Wöhler, "Untersuchungen über das Radikal der Benzoesäure," *Annalen der Pharmacie* 3, no. 3 (1832): 249–282. An 1834 translation by J. C. Booth is reprinted in O. T. Benfey, *From Vital Force to Structural Formulas* (Philadelphia: Beckman Center for the History of Chemistry, 1992), 15.

2. Recipe is from Elisabeth Celnart, *Nouveau manuel complet du parfumeur* (Paris: Roret, 1854), 185–186. The phrase "virtues of plants" is from César Gardeton, *Dictionnaire de la beauté* (Paris: Chez L. Cordier, 1826), 130.

3. Auguste Laurent, "Sur le benzoyle et la benzimide," *Annales de chimie et de physique* [2] 59 (1835): 403. What Laugier and Laurent had isolated was in fact benzil, the dimer of benzoyl, which had the empirical formula as the benzoyl radical but with double the number of atoms.

4. Louis-Bernard Guyton de Morveau, Antoine-Laurent de Lavoisier, Claude-Louis Berthollet, and Antoine Fourcroy, *Méthode de nomenclature chimique* (Paris: Chez Cuchet, 1787), 72.

1. THE STORE OF PROVENCE IN PARIS

1. "no foreign nose can abide," Louis-Sébastien Mercier, *Panorama of Paris: Selections from* Tableau de Paris, ed. Jeremy Popkin (University Park: Pennsylvania State University Press, 1999), 41; "dirty stinking streets," J.-J. Rousseau, *Les confessions,* book 4 (Paris: Seuil, 1967), 181; "shadow and stench," Andrew Hussey, *Paris: The Secret History* (New York: Viking, 2006), 168.

2. Johannes Willms, *Paris, Capital of Europe* (New York: Homes and Meier, 1997), 13.

3. *Almanach du commerce et de toutes les addresses de la ville de Paris, pour l'an VII* (Paris: Chez Favre, 1798). The street is also identified as rue du Bourg-l'Abbé in some sources.

4. Arch. Seine D4B6 carton 42, dos. 2318 (Jean-Michel Miraux), cited in Catherine Lanoë, *La poudre et le fard, une histoire des cosmétiques de la Renaissance aux Lumières* (Seyssel: Champ Vallon, 2008).

5. Jean-François Houbigant opened a shop in 1775 on the rue du Faubourg-Saint-Honoré. Fargeon began working in Paris in 1773 at the rue de Roule. Elisabeth de Feydeau, *A Scented Palace: The Secret History of Marie Antoinette's Perfumer*, trans. Jane Lizop (London: I. B. Tauris, 2006). The Palais Royal itself opened for commerce in 1780, and soon became a common address for luxury perfumers. Natacha Coquery, *Tenir boutique à Paris au XVIIIe siècle. Luxe et demi-luxe* (Paris: Comité des travaux historiques et scientifiques, 2011).

6. Paul Gonnet et al., *Histoire de Grasse et de sa région* (Roanne: Éditions Horvath, 1984), 50; Gabriel Benalloul, Géraud Buffa, Françoise Baussan, Michel Graniou, and Frédéric Pauvaret, *Grasse, l'usine à parfums* (Lyon: Ed. Lieux Dits, 2015); Guy Gilly, *Plantes aromatiques et huiles essentielles à Grasse* (Paris: Harmattan, 1997); Marie-Christine Grasse, ed., *Une Histoire mondiale du parfum: des origines à nos jours* (Paris: Somogy, 2007); Brigitte Bourny-Romagné, *Des épices au parfum: comment les épices ont écrit l'histoire des hommes et des parfums* (Geneva: Aubanel, 2006), 122.

7. Victoria Sherrow, *For Appearance' Sake: The Historical Encyclopedia of Good Looks, Beauty, and Grooming* (Westport, CT: Oryx, 2001); Mike Redwood, *Gloves and Glove-Making* (Oxford: Shire Publications, 2016), 27. Holly Dugan, *The Ephemeral History of Perfume: Scent and Sense in Early Modern England* (Baltimore: Johns Hopkins University Press, 2011), 131.

8. Ghislaine Pillivuyt, *Histoire du parfum: De l'Egypte au XIXe siècle* (Paris: Denoël, 1988), 122.

9. Annick Le Guérer, *Le parfum: Des origines à nos jours* (Paris: Odile Jacob, 2005), 111.

10. Gonnet et al., *Histoire de Grasse*, 127; See also Catherine Lanoë, "Mettre les gants en couleurs: La construction des savoirs des gantiers-parfumeurs, 16e–18e siècles," *Zilsel* 9, no. 2 (2021): 219–236.

11. Joan DeJean, *The Essence of Style: How the French Invented High Fashion, Fine Food, Chic Cafes, Style, Sophistication, and Glamour* (New York: Free Press, 2005), 264.

12. Éliane Perrin and Oliver Buttner, *L'age d'or de la parfumerie à Grasse* (Aix-en-Provence: Édisud, 1996); Jean-Michel Goux, *Grasse au temps des parfumeurs* (St. Martin de la Basque: G. Clergeaud, 2005), 75; Eugénie Briot, "From Industry to Luxury: French Perfume in the Nineteenth Century," *Business History Review* 85, no. 2 (2011): 273–294.

13. Gabriel Benalloul et al., *Grasse, l'usine à parfums* (Lyon: Ed. Lieux Dits, 2015); Liliane Hilaire-Pérez and Catherine Lanoë, "Les savoirs des artisans en France au XVIIIe siècle. Pour une relecture de l'histoire des métiers," in *Mélanges à Daniel Roche*, ed. Dominique Margairaz and Philippe Minard, 345–358 (Paris: Fayard, 2011).

14. Perrin and Buttner, *L'age d'or*, 84.

15. Fabienne Pavia, *The World of Perfume* (New York: Knickerbocker Press, 1996).

16. Louis de Rouvroy, duc de Saint-Simon, *Mémoires complets et authentiques du Duc de Saint-Simon*, vol. 1 (Paris: Hachette, 1856), 319; Catherine Lanoë, *La poudre et le fard: Une histoire des cosmétiques de la Renaissance aux Lumières* (Seyssel: Editions Champ Vallon, 2008); H. Lewis, *The Splendid Century: Life in the France of Louis XIV* (Long Grove, IL: Waveland Press, 1997); Alice Camus, "Le parfumeur Martial: réalité historique du parcours d'un marchand mercier sous Louis XIV," *Bulletin du Centre de recherche du château de Versailles*, 2 October 2020, https://doi.org/10.4000/crcv.18216; Annick Le Guérer, *Le parfum: Des origines à nos jours* (Paris: Odile Jacob, 2005).

17. Alice Camus, "Les savonnettes de Bologne : le 'made in France' d'inspiration italienne," *Nez*, 11 May 2021, https://mag.bynez.com/culture-olfactive/les-savonnettes-de-bologne-le-made-in-france-dinspiration-italienne/.

2. THE ESSENCE OF LIFE

1. For the history of distillation, see Robert J. Forbes, *A Short History of the Art of Distillation* (1948; Leiden: Brill, 1970); Seth Rasmussen, *The Quest for Aqua Vitae: The History and Chemistry of Alcohol from Antiquity to the Middle Ages* (Heidelberg: Springer, 2014); Bruce Moran, *Distilling Knowledge: Alchemy, Chemistry, and the*

Scientific Revolution (Cambridge, MA: Harvard University Press, 2005). For Maria the Jewess, see Naomi Janowitz, *Magic in the Roman World: Pagans, Jews, and Christians* (New York: Routledge, 2001); Raphael Patai, *The Jewish Alchemists: A History and Source Book* (Princeton: Princeton University Press, 1995); for Islamic alchemists, see Salim Ayduz, "Alchemy," *The Oxford Encyclopedia of Philosophy, Science, and Technology in Islam* (Oxford: Oxford University Press, 2014), 136; for the link between perfume and alchemy, see Mandy Aftel, *Essence and Alchemy: A Natural History of Perfume* (London: Bloomsbury, 2001).

2. Forbes, *Distillation*, 89; Rasmussen, *Quest for Aqua Vitae*, 83. See also Nancy Siraisi, *Taddeo Alderotti and His Pupils* (Princeton: Princeton University Press, 1981).

3. Quoted in Rasmussen, *Quest for Aqua Vitae*, 91.

4. Lawrence Principe, *The Secrets of Alchemy* (Chicago: University of Chicago Press, 2013), 129. For more on Paracelsus, see Walter Pagel, *Paracelsus: An Introduction to Philosophical Medicine in the Era of the Renaissance*, 2nd ed. (Basel: Karger, 1982); Charles Webster, *Paracelsus: Medicine, Magic, and Mission at the End of Time* (New Haven: Yale University Press, 2008); Bruce Moran, *Paracelsus: An Alchemical Life* (London: Reaktion Books, 2019).

5. Antoine Baumé, *Chymie expérimentale et raisonnée*, vol. 1 (Paris: Didot Jeune, 1773), 41. The four Galenic humors were phlegm, blood, yellow bile, and black bile. Phlegm contained Aristotle's "watery principle."

6. Ursula Klein and Wolfgang Lefèvre, *Materials in Eighteenth-Century Science: A Historical Ontology* (Cambridge, MA: MIT Press, 2007).

7. Annick Le Guerer, *Scent: The Mysterious and Essential Powers of Smell* (New York: Random House, 1992); Louis Peyron, *Odeurs, parfums et parfumeurs lors des grandes épidémies méridionales de peste Arles 1720* (Arles : Société des Amis du Vieil Arles, 1988); Jean-Alexandre Perras and Erika Wicky, "La sémiology des odeurs au XIXe siècle: du savoir medical à la norme sociale," *Études Françaises* 49, no. 3 (2013): 119–135.

8. Robert Boyle, *Suspicions about Some Hidden Qualities of the Air* (London: W. G., 1674). For more on the history of olfaction, see Alain Corbain, *The Foul and the Fragrant: Odor and the French Social Imagination* (Cambridge, MA: Harvard University Press, 1986); Constance Classen, David Howes, and Anthony Synnott, *Aroma: The Cultural History of Smell* (New York: Routledge, 1994); Chantal Jaquet, *Philosophie de l'odorat* (Paris: Presses Universitaires de France, 2010); Carsten Reinhardt, "The Olfactory Object: Toward a History of Smell in the Twentieth Century," in

Objects of Chemical Inquiry, ed. Ursula Klein and Carsten Reinhardt (Sagamore Beach, MA: Science History Publications, 2014); Mandy Aftel, *Fragrant: The Secret Life of Scent* (New York: Riverhead Books, 2014); Jean-Alexandre Perras and Érika Wicky, eds., *Mediality of Smells / Médialités des odeurs* (Lausanne: Peter Lang, 2021).

9. Hieronymus Brunschwig, *Liber de arte distillandi*, cited in Rasmussen, *Quest for Aqua Vitae*, 102.

10. Marie Meurdrac, *La chymie charitable et facile, en faveur des dames* (1666; Paris: CNRS Editions, 1999).

11. Forbes, *Distillation*, 102.

12. Jean Watin-Augouard, *Il n'y a que Maille . . . Three Centuries of Culinary Tradition* (Versailles: Editions SPSA, 2000). See also Annick Le Guérer, *Quand le parfum portait remède* (Paris: Garde Temps, 2009).

13. *Catalogue général des marchands épiciers, grossiers, droguistes et des marchands épiciers-grossiers-droguistes et des marchands apoticaires-épiciers de cette ville, fauxbourgs et banlieue de Paris* (Paris: Prault, 1773), 39.

14. "Laugier père et fils, Parfumeurs et Distillateurs," Archives de Paris, series 6AZ, folder 684, document 3.

15. Pierre Julien, "Sur les relations entre Macquer et Baumé," *Revue d'histoire de la pharmacie* 80, no. 292 (1992): 65–77; Jonathan Simon, *Chemistry, Pharmacy, and Revolution in France* (New York: Routledge, 2005); E. C. Spary, *Eating the Enlightenment* (Chicago: University of Chicago Press, 2012).

16. Antoine Baumé, *Chymie expérimentale*, vol. 1, ii.

17. Antoine Baumé, *Chymie expérimentale*, vol. 1, iv.

18. Jacques Savary des Brûlons, *Dictionnaire universel de commerce* (Paris: 1723), cited in L. M. Cullen, *The Brandy Trade under the Ancien Régime: Regional Specialisation in the Charente* (Cambridge: Cambridge University Press, 1998), 8, 24.

19. Louis-Sébastien Mercier, *Panorama of Paris: Selections from* Tableau de Paris, ed. Jeremy Popkin (University Park: Pennsylvania State University Press, 1999, 96.

20. Antoine Baumé, *Mémoire sur la meilleure manière de construire les alambics et fourneaux propres à la distillation des vins pour en tirer les eaux-de-vie* (Paris: Didot Jeune, 1778), "disagreeable odor," 23, "contract an odor," 62.

21. Antoine Baumé, *Élémens de pharmacie theorique et pratique* (Paris: Didot Jeune, 1762), 212.

22. Baumé, *Mémoire sur la meilleure manière*, 2.

23. Elisabeth de Feydeau, *A Scented Palace: The Secret History of Marie Antoinette's Perfumer*, trans. Jane Lizop (London: I. B. Tauris, 2006), 21.

24. Lavoisier, "Rapport sur le rouge vegetal," *Oeuvres de Lavoisier*, vol. 4, 224–228 (Paris: Imprimerie Impériale, 1868); see also Catherine Lanoë, "Céruse et cosmétique sous l'ancien régime, XVIe–XVIIIe siècles," in *La céruse: usages et effets Xe–XXe siècles*, ed. Laurence Lestel, Anne-Cécile Lefort, and André Guillerme (Paris: Centre d'histoire des techniques CNAM, 2003).

25. "Annonces et Notices," *Mercure de France* 5 (December 1789).

26. Seymour Mauskopf, "Lavoisier and the Improvement of Gunpowder Production," *Revue d'histoire des sciences* 48, no. 1 (1995): 95–122; Robert Multhauf, "The French Crash Program for Saltpeter Production, 1776–94," *Technology and Culture* 12, no. 2 (1971): 163–181; Charles Gillispie, *Science and Polity at the End of the Old Regime* (Princeton: Princeton University Press, 1980); Patrice Bret, *Annexe III: La Régie des poudres et salpêtres, 1775–1792, Correspondance de Lavoisier*, vol. 5 (Paris: Académie des Sciences, 1993).

27. Six hours a day: Jean-Pierre Poirier, *Lavoisier: Chemist, Biologist, Economist*, trans. Rebecca Balinski (Philadelphia: University of Pennsylvania Press, 1998), 95.

28. H. Thirion, *La vie privée des financiers au XVIIIe siècle* (Paris: Plon, 1896), 431, cited in Poirier, *Lavoisier*, 198.

29. Cullen, *The Brandy Trade*, 69.

30. M. Maçon, "La Ville de Chantilly," *Comptes rendus et mémoires, Société d'histoire et d'archéologie de Senlis* (1912): 1–57, 42.

31. "unhealthy air": J. A. Dulaure, *Réclamation d'un citoyen contre la nouvelle enceinte de Paris, élevée par le fermiers généraux* (no place, no publisher, 1787), cited in Poirier, *Lavoisier*, 172.

32. M. de Lescure, *Correspondance secrete sur Louis XVI*, vol. 1, 579–580, cited in Poirier, *Lavoisier*, 172.

33. Louis-Bernard Guyton de Morveau, Antoine-Laurent de Lavoisier, Claude-Louis Berthollet, and Antoine Fourcroy, *Méthode de nomenclature chimique*

(Paris: Chez Cuchet, 1787), 15. It should be noted that Lavoisier himself did not identify the Aristotelian system as his primary target, but his work did sound the death knell for treating air, water, and fire as elemental. For more on the Chemical Revolution, see Henry Guerlac, *Lavoisier: The Crucial Year* (Ithaca, NY: Cornell University Press, 1961); Frederic Lawrence Holmes, *Antoine Lavoisier: The Next Crucial Year or The Sources of His Quantitative Method in Chemistry* (Princeton, NJ: Princeton University Press, 1998); John G. McEvoy, *The Historiography of the Chemical Revolution: Patterns of Interpretation in the History of Science* (London: Pickering and Chatto, 2010).

34. Guyton de Morveau et al., *Méthode de nomenclature*, 72.

35. Lavoisier, *Traité élémentaire de chimie* (Paris: Chez Cuchet, 1789), 140, cited in Frederic Lawrence Holmes, *Lavoisier and the Chemistry of Life: An Exploration of Scientific Creativity* (Madison: University of Wisconsin Press, 1985), 261. See also Ursula Klein, "Contexts and Limits of Lavoisier's Analytical Plant Chemistry: Plant Materials and Their Classification," *Ambix* 52 (2005): 107–157.

36. Holmes, *Chemistry of Life*, 404.

37. Poirier, *Lavoisier*, 237.

3. REVOLUTION

1. "incommode et chétive," Maximilien Robespierre, *Oeuvres complètes*, vol. 8 (Paris: Aux bureaux de la Revue historiques de la révolution française, 1910), 173.

2. "7 février, 1792": *Archives parlementaires*, 38 (29 January–21 February 1792): 261.

3. Matthew Ramsey, *Professional and Popular Medicine in France 1770–1830: The Social World of Medical Practice* (Cambridge: Cambridge University Press, 1988).

4. Guillotin, *Projet de décret sur l'enseignement et l'exercice de l'art de guérir, présenté au nom du Comité de salubrité* (Paris: Impr. nationale, 1791); Maurice Crosland, "The Officiers de Santé of the French Revolution: A Case Study in the Changing Language of Medicine," *Medical History* 48, no. 2 (April 2004): 229–244; David M. Vess, *Medical Revolution in France, 1789–1796* (Gainesville: University Presses of Florida, 1975).

5. Charles Coulston Gillispie, *Science and Polity in France: The Revolutionary and Napoleonic Years* (Princeton: Princeton University Press, 2004), 52.

6. Georges Vigarello, *Concepts of Cleanliness: Changing Attitudes in France since the Middle Ages,* trans. Jean Birrell (Cambridge: Cambridge University Press, 1988), 158.

7. *Décret de la Faculté de médicine sur les nouveaux bains établis à Paris* (Paris, 1785), cited in Vigarello, *Concepts of Cleanliness,* 158.

8. Goyon de La Plombanie, *L'homme en société,* 2 vols. (Paris: Éditions d'Histoire sociale, 1763), 2: 49–50, cited in Daniel Roche, *The Culture of Clothing: Dress and Fashion in the Ancien Regime,* trans. Jean Birrel (Cambridge: Cambridge University Press, 1989), 393, 389.

9. Lisa DiCaprio, *The Origins of the Welfare State: Women, Work, and the French Revolution* (Urbana: University of Illinois Press, 2007), 120.

10. Thomas Carlyle, *The French Revolution,* vol. 3, book 3 (1837; London: Chapman and Hall, 1896), 116.

11. *Journal of the National Convention,* 27 February 1793. See also Olwen Hufton, *Women and the Limits of Citizenship in the French Revolution* (Toronto: University of Toronto Press, 1992), 27; DiCaprio, *The Origins of the Welfare State,* 120; Albert Mathiez, *La vie chère et le mouvement social sous la Terreur,* vol. 1 (Paris: Payot, 1973).

12. "Séance du Jeudi 27 Juin 1793," *Archives parlementaires* 67 (Paris: Paul Dupont, 1895), 543.

13. Papers of Claire Lacombe, Archives Nationales T 1001 1–3, cited in Mathiez, *La vie chère,* 221.

14. "The Society of Revolutionary Republican Women Joins the Cordeliers to Denounce Traitors," *Women in Revolutionary Paris, 1789–1795: Selected Documents,* trans. and commentary by Darline Gay Levy, Harriet Branson Applewhite, and Mary Durham Johnson (Urbana: University of Illinois Press, 1981), 151.

15. "La Société populaire et Montagnarde de la Commune de Marck District de Calais aux Membres du Comité d'agriculture, 5 messidor l'an ii (23 June 1794)," Archives Nationales, AN F / 12 / 1505 #3134.

16. Charles Thomas Kingzett, *The History, Products, and Processes of the Alkali Trade, Including the Most Recent Improvements* (London: Longmans, Green, 1877), 69; Charles C. Gillispie, "The Discovery of the Leblanc Process," *Isis* 48 (1973): 152–170.

17. Jean-Pierre Poirier, *Lavoisier: Chemist, Biologist, Economist,* trans. Rebecca Balinski (Philadelphia: University of Pennsylvania Press, 1998), 118.

18. Poirier, *Lavoisier,* 243, “it’s about time,” 246.

19. Georges Cuvier, “Éloge historique d’Antoine-François de Fourcroy,” *Recueil des éloges historiques lus dans les séances publiques de l’Institut de France,* vol. 1 (Paris: Librairie de Firmin-Didot frères, fils et Cie, 1861), 299–335, 311.

20. W. A. Smeaton, *Fourcroy, 1755–1809* (Cambridge: Heffer and Sons, 1962), 41.

21. “7 Septembre 1792”: F.-A Aulard, *Recueil des actes du comité de salut public,* vol. 1 (Paris: Imprimerie Nationale, 1889), 48. See also Patrice Bret, “Une administration non révolutionnée? Prosopographie des commissaires au poudres et salpêtres (1775–1817),” in *Nouveaux chantiers d’histoire révolutionnaire: Les institutions et les hommes,* 49–67 (Paris: Comité des travaux historiques et scientifiques, 1995).

22. Smeaton, *Fourcroy,* 42.

23. Gillispie, *Science and Polity in France,* 291–292.

24. Marat, *Les Charlatans modernes* (Paris: Imp. de Marat, 1791), 37.

25. Gillispie, *Science and Polity in France,* 302.

26. Report by Fourcroy to the Convention, 7 Vendémaire, an III (28 September 1794) (Paris: Imprimerie du Comité de salut public, 1794), cited in Jean Dhombres, “Technical Knowledge for Everyone,” *Revolution in Print: The Press in France 1775–1800,* ed. Robert Darnton and Daniel Roche (Berkeley: University of California Press, 1989), 200.

27. Vauquelin and Trusson, *Instruction sur la combustion des végétaux, la fabrication du salin, de la cendre gravelée, et sur la manière de saturer les eaux salpêtries* (1793). Gillispie, *Science and Polity in France,* 389.

28. *Collection générale des décrets rendus par la Convention nationale* (Paris: Baudouin, An II), vol. 16, 272. The National Convention had charged the Commission des subsistances with the question on 20 December 1793. The National Convention had created the Commission des subsistances (National Food Commission) to enforce the general maximum. It operated on direct orders from the Committee of Public Safety.

29. Mémoire sur la manufacture du savon, Archives Nationales, F / 12 / 1505 [no document number].

30. Darcet, Lelièvre, and Pelletier, *Rapport sur la fabrication des savons* (Paris: 12 Nivôse, an III). The committee also examined efforts to produce soda artificially,

surveying nine different processes, including one by Nicolas Leblanc. He was uncooperative, so they published his process against his will and commandeered his factory, although they still could not make enough soda ash. Napoleon returned the factory in 1802, but Leblanc was unable to make a profit and killed himself in 1806. Darcet, Lelièvre, Pelletier, and Giroud, *Description des divers procédes pour extraire la soude du sel marin* (Paris, 1794); Gillispie, "Discovery of the Leblanc Process."

31. "Pelletier aux membres de la commission d'agriculture & arts," Archives Nationales, F / 12 / 1505, #9830.

32. Fourcroy was one of nine professors. Gillispie, *Science and Polity in France*, 395.

33. Order given 31 August 1793. Gillispie, *Science and Polity in France*, 399. On 4 December 1793, the decree of 14 Frimaire an II extended the project to every citizen of France.

34. Gillispie, *Science and Polity in France*, 414.

35. David P. Jordan, *The Revolutionary Career of Maximilien Robespierre* (Chicago: University of Chicago Press, 1989), ix, 275.

36. Morag Martin, *Selling Beauty: Cosmetics, Commerce, and French Society, 1750–1830* (Baltimore: Johns Hopkins University Press, 2009), 49. For perfumers in the Revolution, see also Jean-Alexandre Perras, "The perfumed reaction: The 'petits musqués' of the French Revolution," *Litterature* 185 (2017): 24–38; Elizabeth Amann, *Dandyism in the Age of the Revolution: The Art of the Cut* (Chicago: University of Chicago Press, 2015).

37. On 9 Frimaire, an III, or 29 November 1794, Laugier was nominated to "Décret portant nomination de citoyens pour compléter le comité civil de la section Amis de la Patrie. Du 9 frimaire." *Collection générale des décrets rendus par la Convention nationale*, vol. 28 (Paris: Baudouin, an III), 63.

38. "Affiches, Annonces et Avis Divers," *Journal général de France* (1 June 1793): 2750.

39. Sean Quinlan, "Physical and Moral Regeneration after the Terror: Medical Culture, Sensibility and Family Politics in France, 1784–1804," *Social History* 29, no. 2 (May 2004): 139–164; Nina Rattner Gelbart, "The French Revolution as Medical Event: The Journalistic Gaze," *History of European Ideas* 10, no. 4 (1989): 417–427; E. C. Spary, *Utopia's Garden: French Natural History from Old Regime to Revolution* (Chicago: University of Chicago Press, 2000), 99–102.

40. Mona Ozouf, *Festivals and the French Revolution,* trans. Alan Sheridan (Cambridge, MA: Harvard University Press, 1988), 81.

41. Quinlan, "Physical and Moral Regeneration after the Terror."

42. Simon Schama, *Citizens* (New York: Knopf, 1991), 748.

43. César Gardeton, *Dictionnaire de la beauté* (Paris: Chez L. Cordier, 1826), 130.

44. Gardeton, *Dictionnaire de la beauté, "esprit recteur,"* 129; "amalgam," 130; "regenerator," 132.

45. Gardeton, *Dictionnaire de la beauté,* 132.

46. "Exportations," 14 February 1794, Archives Nationales AF.II.8, #23.

47. Poirier, *Lavoisier,* 319.

48. Poirier, *Lavoisier,* 351.

49. Poirier, *Lavoisier,* 359.

50. Louis Bernard Guyton de Morveau et al., *Encyclopédie méthodique: Chymie, pharmacie et métallurgie,* vol. 6 (Paris: Panckoucke, 1786), 108; Tessier, "Barille," *Dictionnaire des sciences naturelles,* vol 4 (Strasbourg: F. G. Levrault, 1816), 73.

51. Chaptal, *Mes souvenirs sur Napoléon* (Paris: Libraire Plon, 1893), 39.

52. Gillispie, *Science and Polity in France,* 403.

53. Prieur, "Notice sur l'exploitation extraordinaire de salpêtre qui a eu lieu en France pendant les années II et III de la République: ainsi que le nouveau procédé de raffinage de ce sel," *Annales de chimie* 20 (1797): 298–327, cited in André Guillerme, *La naissance de l'industrie à Paris: entre sueurs et vapeurs, 1780–1830* (Seyssel: Champ Vallon, 2007), 70.

54. Gillispie, *Science and Polity in France,* 404.

55. Chaptal, *Mes souvenirs sur Napoléon,* 44. He was placed in charge of what was called the Agence Révolutionnaire des poudres, a body distinct from la Régie des poudres (which was now renamed Agence Nationale des Poudres. On 5 July 1794, the two were combined into one Agence des Poudres et Salpètres).

56. Cuvier did not himself blame Fourcroy but did address the controversy. Cuvier, "Éloge historique d'Antoine-François de Fourcroy," 325.

57. Poirier, *Lavoisier,* 384.

58. J.-A. Chaptal, *Tableau analytique du cours de chymie, fait a Montpellier* (Montpellier: Picot, 1783), 19.

59. Vauquelin and Trusson, *Instruction sur la combustion des végétaux, la fabrication du salin* (Tours: A. Vauquer et Lhéritier, 1794), 23.

60. Prieur, "Sur l'exploitation extraordinaire de Salpêtre, qui a eu lieu en France, pendant les années 2 et 3 de la République," *Annales de chimie* 20 (1797): 298–307, 304.

61. Chaptal, *Mes Souvenirs sur Napoléon,* 55.

62. There are no clear records of how much powder was there. Estimates range from 30 to 150 tons. Thomas Le Roux, "Accidents industriels et regulation des risques: l'explosion de la poudrerie de Grenelle en 1794," *Revue d'histoire modern et contemporaine* 58, no. 3 (2011): 34–62. Thomas Le Roux, "L'explosion de la poudrerie de Grenelle, en 1794," *Santé et Travail* n. 089 (January 2015): 50–51.

63. Chaptal, *Mes souvenirs sur Napoléon,* 46.

64. *The History of Paris from the Earliest Period to the Present Day,* vol. 3 (Paris: A. and W. Galignani, 1832), 314.

65. J.-A. Chaptal, "Observations sur le savon de laine et sur ses usages dans les arts," *Mémoires de l'Institut,* I (an IV / an VI [1798]), lu le 1er prairial, an IV (20 May 1796), cited in Gillispie, *Science and Polity in France,* 488.

4. The Miracle Waters of Cologne

1. For a biography, see Jean Pigeire, *La vie et l'oeuvre de Chaptal (1756–1832)* (Paris: Domat-Montchrestien, 1931); Maurice Crosland, *The Society of Arcueil: A View of French Science at the Time of Napoleon* (Cambridge, MA: Harvard University Press, 1967). *Chaptal: L'itinéraire sans faute de Jean-Antoine Chaptal (1756–1832),* ed. Michel Péronnet, preface by Michel Vovelle (Toulouse: Privat, 1988). For the University of Montpellier, see Elizabeth Williams, *A Cultural History of Medical Vitalism in Enlightenment Montpellier* (New York: Routledge, 2003). For more on vitalism, see Timothy Lenoir, *The Strategy of Life: Teleology and Mechanics in Nineteenth Century German Biology* (Dordrecht: D. Reidel, 1982).

2. Chaptal, *Mes souvenirs sur Napoléon* (Paris: Libraire Plon, 1893), 26.

3. Chaptal, *Discours prononcé à la séance publique de l'École de Santé de Montpellier du premier Brumaire, an V* (Montpellier: Imprimerie de Tournel père et fils, 1796), 6.

4. Chaptal, *Discours*, 5.

5. Chaptal, *Élémens de chimie*, vol. 1 (Montpellier: Jean-François Picot, 1790), xxiii.

6. Chaptal, *Élémens de chimie*, vol. 3 (Montpellier: Jean-François Picot, 1790), 6.

7. Jean-Antoine-Claude Chaptal, *Elements of Chemistry*, vol. 3, trans. William Nicholson (Boston: J. T. Buckingham, 1806) 495.

8. Chaptal, *Chimie appliquée aux arts*, vol. 2 (Paris: Deterville, 1807), 464.

9. Chaptal, *L'art de faire le vin* (Paris: Deterville, 1819), 379. Harry Paul points out that this position is ambiguous in the first edition of 1807, where he refers to it as the *principe doux*, or sweet principle, and the importance of the ferment is only clarified in later editions. Harry Paul, *Science, Wine and Vine* (Cambridge: Cambridge University Press, 1996), 127.

10. Chaptal, *Traité théorique et practique sur la Culture de la Vigne*, vol. 2 (Paris: Chex Delalain, 1801), 14.

11. Fourcroy, *Élémens d'histoire naturelle et de chimie*, vol. 4 (Paris: Chez Cuchet, l'an II), 266.

12. Chaptal, "Observations sur les différences qui existent entre l'acide acéteux et l'acide acétique," *Annales de chimie* 28 (October 1798).

13. H. A. Conner and R. J. Allgeier, "Vinegar: Its History and Development," *Advances in Applied Microbiology* 20 (1976): 82–127.

14. Dickens, *Household Words*, vol. 11 (Leipzig: Bernh. Tauchnitz Jun., 1852), 408.

15. Chaptal, *Traité théorique*, 555.

16. Chaptal, *Élémens de chimie*, 1: 227.

17. Jean Pigeire, *La vie et l'œuvre de Chaptal (1756–1832)* (Paris: Éditions Spès, 1931), 68. For more on Cambacérès, see Jean-Louis Bory, *Les cinq girouettes* (Paris: Ramsay, 1979); Isser Woloch, *Napoleon and His Collaborators* (New York: W. W. Norton, 2002).

18. Woloch, *Napoleon*, 146.

19. Andrew Roberts, *Napoleon: A Life* (New York: Penguin Books, 2015).

20. Chaptal, *Mes souvenirs sur Napoléon*, 55.

21. Antoine Claire Thibaudeau, *Bonaparte and the Consulate*, trans. G. K. Fortescue (London: Methuen, 1908), 12.

22. Chaptal, *Mes souvenirs sur Napoléon*, "force succeeds weakness," 55; 58.

23. Chaptal submitted a report on 9 November 1800 to the Council of State. Quoted in H. C. Barnard, *Education and the French Revolution* (Cambridge: Cambridge University Press, 1969), 200. See also Margaret C. Jacob, *The First Knowledge Economy: Human Capital and the European Economy, 1750–1850* (Cambridge: Cambridge University Press, 2014).

24. Thibaudeau, *Bonaparte and the Consulate*, 304.

25. Barnard calls the Fourcroy Law of 1802 "the last act in the educational history of the Revolution." 209.

26. Fourcroy, "Medicin," *Répertoire universel et raisonné de jurisprudence*, vol. 11 (Paris: J.-P. Roret, 1827), 1.

27. Jonathan Simon and Claude Viel, "Antoine-François de Fourcroy (1755–1809), promoteur de la loi de Germinal an XI," *Revue d'histoire de la pharmacie* 91, no. 339 (2003): 377–394. G. Kersaint, *Antoine François de Fourcroy (1755–1809), sa vie, son oeuvre* (Paris: Editions du Muséum et Centre national de la Recherche scientifique, 1966).

28. Fourcroy's mother was Jeanne Laugier. She had a brother, Jacques-André, who was André Laugier's father. Her sister was Chéradame's wife's mother, making Chéradame and Fourcroy cousins by marriage. André Laugier also married Chéradame's daughter, Jeanne-Marie Pauline. Antoine-Louis Brongniart's mother, Geneviève Fourcroy, was Fourcroy's cousin. Élie Bzoura, "De Chéradame à Fourcroy," *Revue d'histoire de la pharmacie* 57, no. 362 (2009): 119–124, 122.

29. Fourcroy, "Medicin," 1.

30. Matthew Ramsey, *Professional and Popular Medicine in France 1770–1830: The Social World of Medical Practice* (Cambridge: Cambridge University Press, 1988), 179.

31. Ramsey, *Professional and Popular Medicine in France*, 177. See also Matthew Ramsey, "Traditional Medicine and Medical Enlightenment: The Regulation of Secret Remedies in the Ancien Régime," in *La médicalisation de la société française 1770–1830*, ed. Jean-Pierre Goubert (Waterloo, ON: Historical Reflections Press, 1982), 215–232.

32. Madame de Rémusat (Claire Elisabeth Jeanne Gravier de Vergennes), *Memoirs of Madame de Rémusat: 1802–1808*, trans. Cashel Hoey and John Lillie (New York: D. Appleton, 1880), 372. Seizure: Barry Edward O'Meara, *Napoleon in Exile* (Cambridge: Cambridge University Press, 2015), 144. Handkerchief: Frédéric Masson, *Napoléon chez lui: la journée de l'Empereur aux Tuileries* (Paris: E. Dentu, 1894), 102.

33. Jean Vasse, "Napoléon et l'eau de Cologne," *Revue d'histoire de la pharmacie* 57, no. 203 (1969): 497–499.

34. R. H. Horne, *The History of Napoleon* (London: Robert Tyas, 1840), 281.

35. Chaptal, *Mes souvenirs sur Napoléon*, 370.

36. Antoine Henri, *Life of Napoleon*, trans. H. W. Halleck, 4 vols. (New York: D. Van Nostrand, 1864), 1: 70.

37. Masson, *Napoléon chez lui*, "à la Titus," 87; 63.

38. Masson, *Napoléon chez lui*, 82. Shaving himself was highly unusual at the time but kept strange hands with razors away from his throat. Rémusat, *Memoirs*, 372.

39. Masson, *Napoléon chez lui*, 83.

40. Docteur Cabanès, *Dans l'intimité de l'empereur* (Paris: Albin Michel, 1924), 57.

41. Louis Constant Wairy, *Mémoires de Constant, premier valet de chamber de l'empereur sur la vie privée de Napoléon*, 6 vols. (Paris: Chez Ladvocat, 1830), 2: 66.

42. Masson, *Napoléon chez lui*, 89.

43. "Napoleon's Perfumes and Medicines," *Drug Topics*, 21 March 1908, 83.

44. Wilhelm Mönckmeier, *Die Geschichte des Hauses Johann Maria Farina gegenüber dem Jülichs-Platz in Köln* (Berlin: Vowinkel, 1934), 2, 18.

45. Markus Eckstein, *Eau de cologne, Farina's 300th Anniversary* (Cologne: J. P. Bachem Verlag, 2009), 16.

46. Eckstein, *Eau de cologne*, "genuine," 45; 44.

47. Ferdinand Franz Wallraf, cited in Peter Fritzsche, *Stranded in the Present: Modern Time and the Melancholy of History* (Cambridge, MA: Harvard University Press, 2010), 115.

48. Louis François Joseph baron de Bausset-Roquefort, *Private Memoirs of the Court of Napoleon* (Philadelphia: Carey, Lea & Carey, 1828), 299.

49. Constant Wairy, *Mémoires de Constant,* 2: 180.

50. Mönckmeier, *Die Geschichte des Hauses Johann Maria Farina,* 125.

51. L****, *Paris et ses modes ou les soirées parisiennes* (Paris: Chez Michelet, 1803), 191.

52. *Almanach du commerce et de toutes les adresses de la ville de Paris* (Paris: Chez Favre, 1798), 372.

53. L****, *Paris et ses modes,* 191.

54. P. J. Allard, *Almanach de Paris, capitale de l'Empire, et annuaire administratif et statistique du département de la Seine pour l'année 1808* (Paris: Compagnie des Notaire, 1808), 313.

55. "Napoleon's Perfumes and Medicines," *Drug Topics,* vols. 23–24 (1908), 83.

56. "Association entre Jean Marie Farina et Pierre Claude Durochereau pour la fabrication et la vente de l'eau de Cologne, 3 juillet 1806," Archives Nationales, MC / RS / / 948.

57. Charles Jean Auguste Maximilien de Colnet du Ravel, *L'hermite du Faubourg Saint Germain,* (Paris: Chez Pillet ainé, 1825), 1: 202.

58. "Avis—Eau de Cologne," *Affiches, announces et avis divers,* no. 150, 30 May 1807, 2378. The announcement refers to the rue Neuve Saint-Eustache, which was the previous name of rue d'Aboukir.

59. Jeff Horn, *Economic Development in Early Modern France: The Privilege of Liberty, 1650–1820* (Cambridge: Cambridge University Press, 2015), 263.

60. Registre des Dépôts et Actes faits au Greffe du Tribunal de Commerce du département de la Seine, ADS. D16u3.1, Archives de Paris.

61. Mönckmeier, *Die Geschichte des Hauses Johann Maria Farina,* 149.

62. Julien Michel Dufour, "Remèdes secrets," *Répertoire raisonné pour les Préfets, sous-Préfets, Maires . . .* (Paris: Longchamps, 1811), 624; Christian Warolin, "Le remède secret en France jusqu'à son abolition en 1926," *Histoire de la pharmacie* 90, no. 334 (2002): 229–238, 235.

63. Dufour, "Remèdes secrets," 626.

64. Jean-Marie Farina, *Précis sur les propriétés médicales de l'eau de Cologne* (Paris: Warin-Thierry, 1825), 16.

65. Farina, *Précis*, 12.

66. Rapport à S.M. l'Empereur, 11 February 1811, Archives Nationales, AN F.12.2248, #7.

67. Rapport présenté au Ministre de l'Intérieur, 11 February 1811, Archives Nationales, AN F.12.2248, #25.

68. Pierre A. Dubois, "Les Premiers brevets d'invention français concernant la parfumerie alcoolique," *Bulletin de Liaison de la Société Technique des Parfumeurs de France*, no. 2 (January 1978); Liliane Hilaire-Perez, *L'invention technique au siècle des Lumières* (Paris: Albin Michel, 2000).

69. Quesneville, "Eau de Cologne de Laugier, père et fils," *Choix de recettes et formules dans les sciences, les arts, l'économie rurale* (Paris: Imprimerie Dr. Martinet, 1847), 371. Language is from the brevet itself, which was made public in 1824.

70. *À Messieurs les créanciers du sr Antoine-François Laugier, marchand parfumeur* (1828), 3.

5. THE PROBLEM OF VEGETATION

1. "Séance du Lundi, 19 Floréal, an XI," *Procès-verbaux de la Classe de Sciences Mathématiques et Physiques de l'Institut National* (Floréal 19 an XI—9 May 1803), reproduced in Matthieu Gounelle, "The Meteorite Fall at l'Aigle and the Biot Report," in *The History of Meteoritics and Key Meteorite Collections: Fireballs, Falls and Finds*, ed. G. J. H. McCall, A. J. Bowden, and R. J. Howarth (London: Geological Society, 2006): 73–89, 74; see also Simon Schaffer, "Late Enlightenment Crises of Facts: Mesmerism and Meteorites," *Configurations* 26, no. 2 (2018): 119–148.

2. Antoine Lavoisier, "Rapport sur une pierre qu'on prétend être tombée du ciel pendant un orage," in *Oeuvres*, vol. 4 (Paris: Imprimerie impériale, 1868): 40–45.

3. *Journal des debats*, 29 August 1803, cited in Gounelle, "The Meteorite Fall at l'Aigle," 77.

4. Biot, "Une anecdote relative à M. Laplace," *Journal des savants* (February 1850): 65–71, 65–66.

5. Biot, *Relation d'un voyage fait dans le département de l'Orne pour constater la réalité d'un météore observé a l'Aigle le 6 floréal an 11* (Paris: Badouin, 1803), 28.

6. Gounelle, "The Meteorite Fall at l'Aigle."

7. Vauquelin, "Mémoire sur les pierres dites tombées du ciel, "*Annales de chimie* 45 (Year 11): 225–245.

8. Déyeux and Vauquelin, "Observations sur l'état actual de l'analyse végétale, suivies d'une Notice sur l'analyse de plusieurs espèces de sèves d'arbres," *Journal de la société des pharmaciens de Paris* (Year 8 / 1799), 47.

9. William Smeaton, *Fourcroy, Chemist and Revolutionary, 1755–1809* (Cambridge: Heffer, 1962), 167.

10. *Mémoires de l'Institut,* 1806 7 (part 1), 168–222.

11. Smeaton, *Fourcroy*, 145.

12. Fourcroy, "Examen chimique du Cerveau de plusieurs animaux," *Annales de Chimie* 16 (January 1793): 282–322.

13. James Henry Breasted, *The Edwin Smith Surgical Papyrus: published in facsimile and hieroglyphic transliteration with translation and commentary in two volumes* (Chicago: University of Chicago Press, 1991); James P. Allen with an essay by David T. Mininberg, *The Art of Medicine in Ancient Egypt* (New York: Metropolitan Museum of Art, 2005), 13.

14. Berthollet, "Extrait d'un Mémoire sur l'Acide Prussique," *Annales de chimie* 1 (1790): 30–39.

15. Vauquelin, "Expériences qui démontrent la présence de l'acide prussique tout formé dans quelques substances végétales," *Annales de chimie* 45 (Jan 1803): 206–212, 210.

16. Gay-Lussac, "De l'acide prussique," *Annales de chimie* 95 (1815): 136–231, 163.

17. Robiquet, "Considerations sur l'arome," *Annales de chimie* 15 [series 2] (1820): 27–37, 27.

18. Robiquet, "Nouvelles expériences sur l'huile volatile d'amandes amères," *Annales de chimie* 21 [series 2] (1822): 250–254, 253.

19. Martin Booth, *Opium: A History* (New York: Saint Martin's Griffin, 1996), 24.

20. Philip Ball, *The Devil's Doctor: Paracelsus and the World of Renaissance Magic and Science* (New York: Farrar, Straus and Giroux, 2006), 182.

21. Jeff Goldberg and Dean Latimer, *Flowers in the Blood: The Story of Opium* (1981; New York: Skyhorse Publishing, 2014), 51.

22. Jacques-Louis Moreau de la Sarthe, "Opium," *Encyclopédie méthodique,* vol. 11 (Paris: chez Panckoucke, 1824), 153, cited in Roselyne Rey, *The History of Pain,* trans. Louise Elliott Wallace, J. A. Cadden, and S. W. Cadden (Cambridge, MA: Harvard University Press, 1998), 148.

23. Christian Warolin, "The Opiate Pharmacopeia in France from Its Origins to the 19th Century," *Revue d'histoire de la pharmacie* 58, no. 365 (April 2010): 81–90; Patrice Boussel and Henri Bonnemain, *History of Pharmacy and the Pharmaceutical Industry,* trans. Desmond Newell and Frank J. Bové (Paris: Asklepios Press, 1983).

24. Antoine Baumé, *Élémens de pharmacie théorique et pratique* (Paris: Didot Jeune, 1762), 195, 407, 421.

25. "Laugier père et fils, Parfumeurs et Distillateurs," Archives de Paris, series 6AZ, folder 284, document 3.

26. Baumé, *Élémens de pharmacie,* 214.

27. Fourcroy, *Élémens d'histoire naturelle et de chimie,* vol. 1 (Paris: Chez Cuchet, 1789), 21.

28. Chaptal, *Élémens de chimie,* vol. 3 (Montpellier: Jean-François Picot, 1790), 209.

29. Jean Flahaut, "Les Derosne, pharmaciens parisiens, de 1779 à 1855," *Revue d'histoire de la pharmacie* 53, no. 346 (Jan 2005): 221–234.

30. Derosne, "Mémoire sur l'opium: lu à la Société de pharmacie," *Annales de chimie* 45 (1803): 257–286.

31. Derosne, "Mémoire sur l'opium," 281.

32. Jean-Baptiste Sirey, *Jurisprudence du Conseil d'état, depuis 1806,* vol. 4 (Paris: Cour de Harlai, 1818), 288. The rumor had started that he was in prison for embezzling drugs, but this does not seem to be true. Martin Booth, *Opium: A History;* Thomas Dormandy, *Opium: Reality's Dark Dream* (New Haven: Yale University Press, 2012), 115.

33. Louis-Gabriel Michaud, "Armand Séguin," *Biographie universelle ancienne et moderne,* vol. 38 (Paris: Ch. Delagrave, 1843), 669.

34. Séguin, "Sur l'opium," *Annales de chimie et de physique* 92 (1814): 225–245, 228.

35. R. Schmitz, "Friedrich Wilhelm Sertürner and the Discovery of Morphine," *Pharmacy in History* 27, no. 2 (1985): 61–74, 62.

36. "les forces vitals semblaient exaltée": Sertürner, "Analyse de l'Opium.—De la Morphine (nouvel alcali) et de l'Acide méconique, considérés comme parties essentielles de l'opium," *Annales de chimie* 5 (1817): 21–41, 28.

37. "Ueber das Morphium, eine neue salzfähige Grundlage, und die Mekonsäure, als Hauptbestandtheile des Opiums," *Annalen der physik* 55 (1817): 56–90.

38. Sertürner, "Analyse de l'Opium," 22.

39. Sertürner, *Einige Belehrungen für das gebildete und gelehrte Publikum* (Göttingen, 1838), v, cited in Schmitz, "Friedrich Wilhelm Sertürner," 64.

40. Gay-Lussac, "Observation du Rédactuer," following Sertürner, "Analyse de l'opium," *Annales de chimie* (1817): 21–42, 41.

41. Vauquelin, "Examen de l'opium indigène, et réclamation en faveur de M. Séguin, de la découverte de la morphine et de l'acide méconique," *Annales de chimie* 9 (1818): 282.

42. John E. Lesch, "Conceptual Change in an Empirical Science: The Discovery of the First Alkaloids," *Historical Studies in the Physical Sciences* 11, no. 2 (1981): 305–328; Sacha Tomic, *Aux origines de la chimie organique: méthodes et pratiques des pharmaciens et des chimistes (1785–1835)* (Rennes: Presses Universitaires de Rennes, 2010).

43. Pierre-Joseph Pelletier and Joseph-Bienaimé Caventou, "Mémoire sur un nouvel alcali végétal (la Strychnine)," *Annales de chimie* 10 (1819): 142–177.

44. Pelletier and Caventou, "Mémoire sur un nouvel alcali," 143.

45. Pelletier and Caventou, "Mémoire sur un nouvel alcali," 146.

46. Pelletier and Caventou, "Suite: Des recherches chimiques sur les quinquinas," *Annales de chimie* 15 (1820): 337–365.

6. A Temple of Industry

1. *Notice historique sur l'ancien prieure de Saint-Martin des Champs et sur le Conservatoire national des arts et metiers* (Paris, 1882), 14.

2. *Notice historique*, 44.

3. Robert J. Forbes, *A Short History of the Art of Distillation* (1948; Leiden: Brill, 1970), 309.

4. "Annonces Générales," *Le constitutionnel* 7 April 1824.

5. César Gardeton, *Dictionnaire de la beauté* (Paris: Chez L. Cordier, 1826), "spirituous," 137, "*sucs*, " 136, "action," 137.

6. Gardeton, *Dictionnaire de la beauté*, 135.

7. Antonio García Belmar and José Ramón Bertomeu-Sánchez, "Louis Jacques Thenard's Chemistry Courses at the College de France, 1804–1835," *Ambix* 57, no. 1 (March 2010): 48–63; Sacha Tomic, *Aux origines de la chimie organique: méthodes et pratiques des pharmaciens et des chimistes (1785–1835)* (Rennes: Presses Universitaires de Rennes, 2010).

8. Robert Fox, *The Savant and the State: Science and Cultural Politics in Nineteenth-Century France* (Baltimore: Johns Hopkins University Press, 2012), 27; Anne-Claire Déré and Gérard Emptoz, *Autour du chimiste Louis-Jacques Thénard (1777–1857). Grandeur et fragilité d'une famille de notables au XIXe siècle* (Chalon-sur-Saône: Université pour Tous de Bourgogne, 2008), 242–263.

9. Belmar and Bertomeu-Sánchez, "Louis Jacques Thenard's Chemistry Courses," 55.

10. Louis-Réné Le Canu, *Souvenirs de M. Thénard* (Paris: Typographie de Mme Ve Dondey-Dupré, 1857), 45.

11. Félix Boudet, José Luis Casaseca, Louis Serbat, and Antoine Boissenot. Belmar and Bertomeu-Sánchez, "Louis Jacques Thenard's Chemistry Courses."

12. Justus Liebig, "À M. le Baron Thénard," *Chimie organique appliquée à la physiologie végetale et à l'agriculture*, trans. Charles Gerhardt (Paris: Chez Fortin, 1841), dedication, n.p., 2nd page after title page.

13. August Hofmann, *Zur Erinnerung an Jean Baptiste André Dumas* (Berlin: Ferd. Dûmmler's Verlagsbuchhandlung, 1885), 6. For more on Dumas, see Marc Tiffeneau, *Jean-Baptiste Dumas (1800–1884)* (Paris: Laboratoires G. Beytout, 1934); L. Klosterman, "A Research School of Chemistry in the Nineteenth Century," *Annals of Science* 42 (1985): 1–80; M. Chaigneau, *Jean-Baptiste Dumas, chimiste et homme politique* (Paris: Guy le Prat, 1984); Alan Rocke, *Nationalizing Science: Adolphe Wurtz and the Battle for French Chemistry* (Cambridge, MA: MIT Press, 2000); Jimmy Drulhon, *Jean-Baptiste Dumas (1800–1884)—La vie d'un chimiste dans les allées de la science et du pouvoir* (Paris: Éditions Hermann, 2011).

14. Rocke, *Nationalizing Science*, 25

15. "Paris, 27 October," *Le Messager des chambres*, no. 241 (October 28, 1828): 3.

16. Édouard Laugier and Anton de Kramer, *Tableaux synoptiques, ou Abrégés des caractères chimiques des bases salifiables* (Paris: Dondey-Dupré, 1828).

17. Anselme Payen, *Traité élémentaire des réactifs, leurs préparations, leurs emplois . . .*, vol. 1 (1830).

18. Édouard Laugier, *Nomenclature chimique* (Paris: impr. de Dondey-Dupré, no date).

19. Italics in original. Édouard Laugier, "Sur la résine d'indigo et le tannin artificiel; par le Dr. Buff.," *Bulletin universel des sciences et de l'industrie* 12 (1829): 284–285.

20. Chevreul and Serrulas, "De l'influence des substances organique sur les caractères chimiques des sels minéraux," Séance du 12 Avril, in *Mémorial des Hopitaux du Midi*, Prof. Delpach, ed., (Paris: Chez Gabon, 1830), vol. 2, 251.

21. Masha Belenky, *Engine of Modernity: The Omnibus and Urban Culture in Nineteenth-century Paris* (Manchester: University of Manchester Press, 2019).

22. Wilhelm Hauff, "Freie Stunden am Fenster," *Phantasien und Skizzen* (Stuttgart: Gebrüder Franckh, 1828), 129.

23. Élisabeth de Feydeau, *Le roman des Guerlains* (Paris: Flammarion, 2017).

24. Édouard Laugier, "Savons de toilette," *Dictionnaire technologique*, vol. 19 (Paris: Thomine, 1831), 177.

25. Peter Hervé, *The New Picture of Paris from the Latest Observations* (London: Sherwood, Gilbert and Piper, 1829), 525.

26. Antoine Nicolas Béraud and Pierre Joseph Spiridion Dufey, *Dictionnaire historique de Paris*, vol. 2 (Paris: Librairie Nationale et Étrangère, 1825), 70.

27. Émery, Mèlier, and Guibourt, "Note sur les eaux distillées de fleurs d'oranger, de roses, etc., et sur la présence des sels métalliques de cuivre et de plomb dans ces liquides," *Bulletin de l'Académie royale de médecine* 11 (1845–1846): 583–591, 584.

28. Émery, Mèlier, and Guibourt, "Note sur les eaux distillées de fleurs d'oranger," 584.

29. Labarraque and Pelletier, "Rapport fait au conseil de salubrité sur un sel de plomb contenu dans l'eau de fleurs d'oranger, Paris, le 25 sept. 1829," *Annales d'hygiène publique et de médecine légale*, vol. 4 (Paris: E. Crochard, 1830): 55–60, 56.

30. John Pannabecker, "Research, Invent, Improve: A Dictionnaire Technologique for Non-elites (1822–35)," *Technology and Culture* 59, no. 3 (July 2018): 546–589.

31. Lenormand, "Discours Préliminaire, " *Dictionnaire technologique*, quoted in Pannabecker, "Research, Invent, Improve," 546–547.

32. Édouard Laugier, "Parfum," *Dictionnaire technologique, ou nouveau dictionnaire universel des arts et métiers, et de l'economie industrielle et commerciale*, vol. 15 (Paris: Thomine, 1829): 325–347.

33. Laugier, "Parfum," "natal soil," 326, "profit," 347.

34. Laugier, "Parfum," "eaux d'odeurs," 334; 336.

35. Joséphine Lebassu, "Mme Bayle-Celnart," *Biographie des femmes-auteurs*, vol. 1 (Paris: Armand-Aubrée, 1836), 202.

36. Lebassu, "Mme Bayle-Celnart," 211.

37. Madame Celnart, *Manuel des dames l'art de l'elegance*, 2nd ed. (Paris: Roret, 1833), "*toilette*, " 1, "plague," 92.

38. Celnart, *Manuel des dames*, "borrow," 76, "procure," 95.

39. Celnart, *Nouveau Manuel complet du parfumeur* (Paris: Roret, 1845), vi.

40. Campbell Morfit, *Perfumery: Its Manufacture and Use* (Philadelphia: Carey and Hunt, 1847).

41. "Lycée commercial et industriel," *Journal de commerce*, 8 December 1828, NP.

42. Armand Marrast, "Méthode Jacotot," *Journal de la langue française* 5 (1831): 60.

43. *Le Constitutionnel*, 2 November 1828.

44. *Le Courrier*, 8 September 1827, 4.

45. "Lycée commercial et industriel," *Le Courrier*, 12 October 1828, 4.

46. Quoted in William Brock, *Justus von Liebig: The Chemical Gatekeeper* (Cambridge: Cambridge University Press, 2002), 43.

47. Brock, *Justus von Liebig*, 44.

48. C. de Camberousse, *Histoire de l'École Centrale des Arts de Manufactures* (Gauthier-Villars, 1879), 22.

49. Aaron Ihde, *The Development of Modern Chemistry* (New York: Harper and Row, 1964), 193; J. R. Partington, *A Short History of Chemistry* (New York: Macmillan, 1957), 240.

50. Camberousse, *Histoire de l'École Centrale*; page 32 has a plan showing every room.

51. Camberousse, *Histoire de l'École Centrale*; see also J. Weiss, *The Making of Technological Man* (Cambridge, MA: MIT Press, 1982).

52. Théodore Olivier and Eugène Péclet, *École Centrale des Arts et Manufactures* (Paris: Chez Béchet Jeune, 1829), 2.

53. Marcel Chaigneau, *Jean-Baptiste Dumas: Sa vie, son oeuvre, 1800–1884* (Paris: Guy le Prat, 1984), 91.

7. LOST ILLUSIONS

1. Mark Traugott has a map of the barricades of 1830, showing a dense collection along the rue Bourg-l'Abbé. Mark Traugott, *The Insurgent Barricade* (Berkeley: University of California Press, 2010), Map 2A.

2. William Hone, *Full Annals of the Revolution in France, 1830* (London: T. Tegg, 1830), 55.

3. "Extrait des minutes du Greffe de la Justice de Paix du sixième arrondissement de la Ville de Paris, Faillite Laugier," Archives de Paris, D11u3.

4. "Rapport des syndics, Faillite Laugier," Archives de Paris, D11u3.

5. Honoré de Balzac, *Traité de la vie élégante* (Paris: Librarie Nouvelle, 1854), 79.

6. Honoré de Balzac, *Traité de la vie élégante*, annotated by Marie-Christine Natta (1854; Clermont-Ferrand: Presses Universitaires Blaise Pascal, 2000), 47.

7. "Faillites, Jugemens du 12 août, 1830," *Gazette des Tribuneaux*, no. 1558, 14 August 1830, 940.

8. Balzac, *La Comédie humaine*, ed. Pierre-Georges Castex, 12 vols. (Paris: Gallimard, 1977), 6: 81.

9. Madeleine Fargeaud, "Balzac, le commerce et la publicité," *L'Année balzacienne* (1974): 187–198.

10. *Bazar parisien*, 1826, 488. Linzy Erika Dickinson proposed an alternative source of inspiration: a catalog Balzac had printed from the pharmacists Dissey and

Pivier. Linzy Erika Dickinson, *Theatre in Balzac's La Comédie Humaine* (Amsterdam: Rodopi, 1994), 293.

11. S. L. Kotar and J. E. Gessler, *Cholera: A Worldwide History* (Jefferson, NC: McFarland, 2014), 89. See also Catherine Kudlick, *Cholera in Post-Revolutionary Paris: A Cultural History* (Berkeley: University of California Press, 1996); François Delaporte, *Disease and Civilization: The Cholera in Paris, 1832* (Cambridge, MA: MIT Press, 1986).

12. *Instruction populaire sur les principaux moyens à employer pour se garantir du choléra-morbus et sur la conduit à tenir lorsque cette maladie se declare* (Grenoble: Imprimerie de F. Allier, 1832), 2.

13. D. B. J. L. Millot, *Histoire pharmacologique du camphre* (Paris: Mansut, 1837), 31.

14. Toussaint Rapou, *Seul traitement préservatif et curatif du choléra asiatique* (Paris: Baillière, 1831), 32; François Gabriel Boisseau, *Traité du choléra-morbus considéré sous le rapport médical et administratif* (Paris: Chez J.-B. Baillière, 1832).

15. Millot, *Histoire pharmacologique du camphre*, 2.

16. "épidémie de choléra, suspension de l'enseignement," (1832), Le fonds d'archives de l'École Centrale des Arts et Manufactures, 20170270 / 8, Archives Nationales.

17. M. Arezula, "Extrait d'une dissertation de M. Proust Qui a pour titre, Résultat des expériences faites sur le camphre de Murcie," *Annales de chimie* 4 (1790): 179–209.

18. Thénard, "Essai sur la combinaison des acides avec les substances végétales et animales, " *Mémoires de physique et de chimie, de la Société d'Arcueil* 2 (1809): 23–41.

19. W. H. Brock, *Justus von Liebig: The Chemical Gatekeeper* (Cambridge: Cambridge University Press, 1997); Alan J. Rocke, *Nationalizing Science: Adolphe Wurtz and the Battle for French Chemistry* (Cambridge, MA: MIT Press, 2001); Catherine M. Jackson, "The 'Wonderful Properties of Glass': Liebig's Kaliapparat and the Practice of Chemistry in Glass," *Isis* 106, no. 1 (2015): 43–69; Melvyn C. Usselman, Christina Reinhart, Kelly Foulser, and Alan J. Rocke, "Restaging Liebig: A Study in the Replication of Experiments," *Annals of Science* 62, no. 1 (2005): 1–55.

20. J. Liebig, "Sur un nouvel appareil pour l'analyse des substance organiques; et sur la composition de quelques-unes de ces substances," *Annales de chimie et de physique* 47 (1831): 147–197. Rocke, *Nationalizing Science*, 43.

21. "Lettre de M. Dumas à M. Gay-Lussac, sur les procédés de l'analyse organique," *Annales de chimie et de physique* 47 (1831): 198–213, 212.

22. Rocke, *Nationalizing Science*, 46; Jackson, "Wonderful Properties of Glass."

23. Dumas, "Note de M. Dumas sur diverses combinaison de l'hydrogène carboné," *Annales de chimie et de physique* 48 (1831): 430–432, 432; also J. Liebig, "Sur la composition de l'acide camphorique et du camphre," *Annales de chimie et de physique* 47 (1831): 95–101.

24. Jean-Baptiste Dumas, "Mémoire sur les substances végétales qui se rapprochent du camphre, et sur quelques huiles essentielles," *Annales de chimie et de physique* 50 (1832): 225–240, 228.

25. Dumas, "Mémoire sur les substances végétales," 229.

26. Ashbel Green, "Extract of a letter, dated Paris, 20th May," *Christian Advocate*, vol. 10 (Philadelphia: A. Finley, 1832), 315.

27. "Bulletin des séances de l'Academie royale des Sciences: Séance du lundi 11 juin," *Annales de chimie et de physique* 50 (1832): 442. On June 18, he gave another "very favorable" report (443).

28. Dumas, "Mémoire sur les substance végétales." Dumas, "Sur les camphres artificiels des essences de térébenthine et de citron," *Annales de chimie et de physique* 52 (1833): 400.

29. "Luminous Plants," *Hardwicke's Science-gossip* 10 (1872): 121–125, 122. Dr. Hahn, "Inflammability of the Flowers of Dictamnus Albus," *Journal of Botany, British and Foreign*, vol. 1 (London: Robert Hardwicke, 1863), 345.

30. Biot, "Sur l'Inflammation de la Fraxinelle (*dictamus alba*)," *Annales de chimie et de physique* 50 (1832): 386.

31. Louis Graves, *Précis statistique sur le canton de Liancourt, arrondissement de Clermont* (Oise, 1837), 76; Jacques Cambry, *Description du département de l'Oise*, vol. 1 (Paris: P. Didot, 1803), 291.

32. Maurice Crosland, *Gay-Lussac: Scientist and Bourgeois* (Cambridge: Cambridge University Press, 1978), 29; Charles Coulston Gillispie, *Science and Polity in France: The Revolutionary and Napoleonic Years* (Princeton: Princeton University Press, 2004), 639.

33. Ken Alder, *The Measure of All Things* (New York: Free Press, 2003).

34. Maurice Crosland, *Society of Arcueil: A View of French Science at the Time of Napoleon I* (Cambridge, MA: Harvard University Press, 1967); Robert Fox, "The Rise and Fall of Laplacian Physics," *Historical Studies in the Physical Sciences* 4 (1974): 89–136.

35. Isaac Newton, Query 26, *Opticks* (London: William Innys, 1720), 333.

36. Biot, "Phénomènes de polarization successive, observés dans des fluides homogènes," *Bulletin des Sciences, par la Société Philomatique de Paris* 1 (1815): 190–192.

37. Roger Hahn, *Pierre Simon Laplace, 1749–1827: A Determined Scientist* (Cambridge, MA: Harvard University Press, 2005), 131.

38. Biot, "Mémoire sur les rotations que certaines substances impriment aux axes de polarisation des rayons lumineux," *Mémoires de l'Académie royale des sciences de l'Institut de France* 2 (1817): 41–136, 131.

39. Frédéric Leclercq, "Arago, Biot et Fresnel expliquent la polarisation rotatoire," *Revue d'histoire des sciences* 66 (2013 / 2): 395–416.

40. The attendance records at the Academy of Sciences show that he appeared only sporadically in the winter and almost not at all in the months from March to November. "Tableau de Présence," *Procès-Verbaux des Séances de l'Académie tenues depuis la fondation de l'Institut jusqu'au mois d'août 1835 Publiés conformént à une décsion de l'Académie par M.M. les Secrétaires Perpétuels* 4 (1916): 744.

41. "Extrait d'une lettre de M. Biot sur les transformations opérées par la vie végétale dans les produits carbonisés qui servent d'aliment aux jeunes individus. Lu à l'Academie des Sciences de Paris . . . par M. Becquerel." Letter from Nointel, 11 May 1833, *Nouvelles annales du Muséum d'histoire naturelle,* II (1833): 365.

42. Emile Picard, "La vie et l'œuvre de Jean-Baptiste Biot," *Éloges et discours académiques* (Paris: Gauthiers-Villars, 1931), 261.

43. Biot, "Mémoire sur la polarisation circulaire et sur ses application à la chimie organique, lu le 5 novembre, 1832," *Mémoires de l'Académie des Sciences* 13 (1835): 39–175, 140.

44. Biot, "Mémoire sur la polarisation circulaire."

45. Biot, "Mémoire sur la polarisation circulaire," recently published, 146, counterclockwise, 149, "indubitable," 155. Biot also got some of the naphthalene

prepared by Laurent, calling it "analogous to essential oils by the nature of its chemical elements" and noting that it acted in the same direction as turpentine, but with four times less energy. It was long assumed Biot had made a mistake, and naphthalene was not optically active, but more than fifty years later he was proved right.

46. Dumas, "Rapport sur un Mémoire de M. Payen, relatif à l'analyse élémentaire de l'amidon et à celle de la dextrine," *Compte rendus de l'Académie des Sciences* 5 (1837): 898–905, 903.

47. Dumas, "Rapport sur un Mémoire de M. Payen," 898.

48. For biographical information on Laurent, see Jean Jacques, "Liste chronologique des publications d'Auguste Laurent," *Archives de l'Institut Grand-Ducal de Luxembourg* 22 (1955); Robert Stumper, "La vie et l'oeuvre d'un grand chimiste, pionnier de la doctrine atomique: Augustin Laurent, 1807–1853," *Archives de l'Institut grand-ducal de Luxembourg, section des sciences,* n.s. 20 (1951–1953): 47–93; Clara deMilt, "Auguste Laurent, Founder of Modern Organic Chemistry," *Chymia* 4 (1953): 85–114; Édouard Grimaux, "Auguste Laurent," *Revue scientifique* [4] 6 (1896); Marya Eunice Novitski, *August Laurent and the Prehistory of Valence* (Chur, Switzerland: Harwood Academic, 1992); Marika Blondel-Mégrelis, *Dire les Choses—Auguste Laurent et la méthode chimique* (Paris: J. Vrin, 1996).

49. Dumas, "Mémoire sur les Substances végétales." Dumas and Laurent had worked together on "Recherches sur les combinaisons de l'hydrogène et du carbone," but only Dumas's name was on the publication.

50. Lissa Roberts and Joppe van Driel, "The Case of Coal," in *Compound Histories: Materials, Governance and Production, 1760–1840,* ed. Lissa Roberts and Simon Werrett (Leiden: Brill, 2018); Louise Lyle and David McCallam, eds., *Histoires de la Terre: Earth Sciences and French Culture 1740–1940* (Amsterdam: Rodopi, 2008); Barbara Freese, *Coal: A Human History* (New York: Basic Books, 2003).

51. Martin Rudwick, *The Meaning of Fossils: Episodes in the History of Palaeontology,* 2nd ed. (Chicago: University of Chicago Press, 1985).

52. Alexandre Brongniart, "On Fossil Vegetables Traversing the Beds of Coal Measures, (Annales des Mines, April 1821)," in *A Selection of Geological Memoirs Contained in the Annales des Mines,* trans. Henry Thomas de la Beche (London: William Phillips, 1836), 208–209.

53. Dumas, *Traité de chimie appliquée aux arts,* vol. 1 (Paris: Chex Béchet Jeune, 1828), 598.

54. John Kidd, "Observations on Naphthalene, a peculiar substance resembling a concrete essential oil, which is produced during the decomposition of coal tar, by exposure to a red heat," *Philosophical Transactions* 111 (1821): 209–221.

55. Michael Faraday, "On new compounds of carbon and hydrogen, and on certain other products obtained during the decomposition of oil by heat," *Philosophical Transactions* 115 (1825): 440–466.

56. Laurent, "Sur un nouveau moyen de préparer la naphtaline et sur son analyse," *Annales de chimie et de physique* 49 (1832): 214–221.

57. All quotations from Laurent, "Sur un nouveau moyen," 217.

58. Satish Kapoor, "The Origins of Laurent's Organic Classification," *Isis* 60, no. 4 (1969): 477–527, 478.

59. Kapoor, "The Origins of Laurent's Organic Classification," 479. Rocke, *Nationalizing Science*; Seymour Mauskopf, "Crystals and Compounds," *Transactions of the American Philosophical Society*, n.s. 66, part 3 (1976): 1–82.

60. Novitski, *Auguste Laurent and the Prehistory of Valence*, 32.

61. Dumas, "Réponse de M. Dumas a la lettre de M. Berzelius, " *Compte rendus de l'Académie des Sciences* 6 (1838): 689–702, 697.

62. J. Jacques, "Auguste Laurent et J.-B. Dumas d'après une correspondance inédite," *Revue d'histoire des sciences* 6 (1953): 329–349, 336.

8. RADICALS AND BOHEMIANS

1. "Laurent to Dumas, 12 June 1836," cited in Marcel Chaigneau, *Jean-Baptiste Dumas: sa vie, son oeuvre, 1800–1884* (Paris: Guy Le Prat, 1984), 137.

2. Balzac, "The Atheist's Mass," *Comédie Humaine*, trans. Ellen Marriage (Philadelphia: Gebbie, 1898), 371.

3. Robert Stumper, "La vie et l'œuvre d'un grand chimiste, pionnier de la doctrine atomique: Augustin Laurent," *Archives de l'Institut grand-ducal de Luxembourg, section des sciences*, n.s. 20 (1953): 47–93, 52.

4. David Albert Griffiths, *Jean Reynaud, encyclopédiste de l'époque romantique, d'après sa correspondance inédite* (Paris: M. Rivière, 1965); Henri Martin, *Jean Reynaud* (Paris: Furne et Cie, 1863), 7. See also John Tresch, *The Romantic Machine: Utopian Science and Technology after Napoleon* (Chicago: University of Chicago Press, 2012).

5. Martin, *Jean Reynaud*, 6.

6. Jean Jacques, "Auguste Laurent (1807–1853), collaborateur de l'Encyclopédie nouvelle (1836–1841)," *Compte rendus de l'Académie des Sciences* 324, series 2b (1997): 197–200. Only four of the eight volumes have tables with the authors' names. In them, Laurent is listed as the author of five articles: "Chimie," "Chalameau," "Cobalt," "Combinaison," and "Combustion." Jean Jacques has also attributed several articles without authors to Laurent, including with virtual certainty "Cristallographie" and "Fermentation," and with high probability "Dimorphisme," "Diamant," and "Étain."

7. Auguste Laurent, "Chimie," *l'Encyclopédie nouvelle*, ed. Pierre Leroux and Jean Reynaud, vol. 3 (Geneva, 1836–1842), 520–524, 520.

8. "Fermentation," *l'Encyclopédie nouvelle*, ed. Pierre Leroux and Jean Reynaud, vol. 5 (Paris: Librairie de Charles Gosselin, 1843), 276–282, 276.

9. Auguste Laurent, "Sur de nouveaux chlorures et brômures d'hydrogène carboné," *Annales de chimie et de physique* [2] 59 (1835): 198–199.

10. Auguste Laurent, "Sur la nitronaphtalase, la nitronaphtalèse et la naphtalase," *Annales de chimie et de physique* 59 (1835): 376–397.

11. Auguste Laurent, "Sur la chlorophénise et les acides chlorophénisique et chlorophénèsique," *Annales de chimie et de physique* 63 (1836): 27–44, 30.

12. Laurent, "Sur la nitronaphtalase," 388.

13. "À M. Laugier, en témoignage de la reconnaissance de l'auteur qui était peu *chimiste*." M. G. Barral, "Quel est ce Laugier?" *La chronique médicale* 14 (1907): 405–407; Madeleine Fargeaud, *Balzac et "la Recherche de l'absolu"* (Paris: Hachette, 1968).

14. Charles Dupin, "Avant propos," *Rapport du jury central sur les produits de l'industrie française exposés en 1834*, vol. 1 (Paris: Imprimerie royale, 1836), xxi.

15. Dupin, *Rapport*, 1: 5.

16. *Catalogue des produits de l'industrie française admis à l'exposition publique sur la place de la Concorde en 1834* (Paris: Pihan-Laforest, imprimeur, 1834), iv.

17. *Notice des produits de l'industrie française* (Paris: Everat, 1834), xi.

18. *Journal des connaissances usuelles et pratiques*, issues 106–107 (1834), 85.

19. Charles Dupin, *Rapport du jury central sur les produits de l'industrie française exposés en 1834*, vol. 3 (1836), 351.

20. Baumé, *Élémens de pharmacie theorique et pratique* (Paris: Didot Jeune, 1762), 551.

21. "Appareil distillatoire dit appareil alcoométrique à distillation et rectification sans fin. B. d'inv de 10 ans, délivré le 8 octobre 1836, à Laugier, professeur de chimie, à paris, rue Bourg-l'Abbé, no. 41," *Catalogue des brevets d'invention, d'importation et de perfectionnement délivrés depuis le 1er janvier 1828 jusqu'au 31 décembre 1842* (Paris: Bouchard-Huzard, 1843), 166.

22. "Revue scientifique," *L'illustration* 25 (1855): 363.

23. Nicolas Basset, *Guide theorique et pratique du fabricant d'alcools et du distillateur*, vol. 3 (Paris: Dictionnaire des Arts et Manufactures, 1873), 219.

24. W. H. Brock, *Justus von Liebig: The Chemical Gatekeeper* (Cambridge: Cambridge University Press, 1997), 57. See also J. B. Morrell, "The Chemist Breeders," *Ambix* 9 (1972): 1–46.

25. Brock, *Justus von Liebig*, 22.

26. Pettenkofer, 1877, cited in Brock, *Justus von Liebig*, 51.

27. Catherine Jackson, "Analysis and Synthesis in Nineteenth-Century Organic Chemistry" (PhD diss., University of London, 2008), 32.

28. Robiquet et Boutron-Charland, "Nouvelles expériences sur les amandes amères, et sur l'huile volatile qu'elles fournissent." *Annales de chimie et physique* 44 (1830): 352–382, "true composition," 352, "crystalline substance," 376.

29. Nostradamus, *Excellent & moult utile opuscule à touts necessaire, qui desirent avoir cognoissance de plusieurs exquises receptes* (Lyon: A. Volant, 1555), 43.

30. Fourcroy and Vauquelin, "Second mémoire pour servir à l'histoire naturelle chimique et médicale de l'urine humaine," *Annales de chimie* 32 (an 8 / 1799): 80–113; Fourcroy and Vauquelin, "Sur une substance de l'ile de Caprée, analysée par M. Laugier," *Annales de chimie* 66, series 1 (1808): 104–112.

31. Robiquet et Boutron-Charland, "Nouvelles expériences sur les amandes amères," 353.

32. Brock, *Justus von Liebig*, 77.

33. *Liebig's und Wöhler's Briefwechsel,* ed. August Hofmann, vol. 1 (Braunschweig: F. Viewig und Sohn, 1888), 53–54, cited in Brock, *Justus von Liebig,* 78.

34. J. Liebig and F. Wöhler, "Untersuchungen über das Radikal der Benzoesäure," *Annalen der Pharmacie* 3, no. 3 (1832): 249–282. An 1834 translation by J. C. Booth is reprinted in O. T. Benfey, *From Vital Force to Structural Formulas* (Philadelphia: Chemical Heritage Foundation, 1992), 35.

35. This was a letter from Berzelius to Liebig and Wöhler, dated 2 September 1832, printed at the end of their memoir. *Ann.* 3 (1832): 282–286; *Ann. Phys.* 26 (1832): 480–485, cited in J. R. Partington, *A Short History of Chemistry* (New York: Macmillan, 1937), 328.

36. Liebig and Wöhler, "Untersuchungen über das Radikal der Benzoesäure," 15.

37. Frederic Holmes, "Justus Liebig and the Construction of Organic Chemistry," in *Chemical Sciences in the Modern World,* ed. Seymour Mauskopf (Philadelphia: University of Pennsylvania Press, 1994), 130.

38. Liebig to Berzelius, 4 August 1831, in *Berzelius and Liebig: Ihre Briefe von 1831–45,* ed. Justus Carrière (Munich: J. F. Lehmann, 1893): 13–16, cited in Hans-Werner Schütt, *Mitscherlich, Prince of Prussian Chemistry,* trans. William E. Russey (Philadelphia: Chemical Heritage Foundation, 1997), 190.

39. Liebig to Wöhler, 1 May 1832, in Schütt, *Mitscherlich,* 191.

40. Mitscherlich, "Über das Benzin und die Säuren der Öl- and Talgarten," *Annalen* 9 (1834): 39–48, see also Schütt, *Mitscherlich,* 140.

41. Justus von Liebig, Addendum, in Mitscherlich, "Über das Benzin," 56, cited in Schütt, *Mitscherlich,* 144.

42. Mitscherlich, "Über das Benzin," 56, cited in Schütt, *Mitscherlich,* 144.

43. Liebig to Berzelius, 25 March 1834, *Berzelius and Liebig, Ihre Breife,* vol. 1831–45, 84, cited in Schütt, *Mitscherlich,* 144.

44. Liebig to Wöhler, 8 March 1834, *Liebig and Wöhler, Briefwechsel,* vol. 1, 79–81, cited in Schütt, *Mitscherlich,* 144.

45. Wöhler to Liebig, 3 March 1834, in *Liebig and Wöhler, Briefwechsel,* vol. 1, 79, cited in Schütt, *Mitscherlich,* 192.

46. "Annual Report, Read by the President Philip Yorke, Anniversary Meeting of March 30, 1854" ("President's Address"), *Quarterly Journal of the Chemical Society of London* 7, no. 26 (1855): 144–159, 152.

47. "Savons de toilette," *Dictionnaire universel des arts et métiers et de l'économie industrielle et commerciale* (Paris: Chez le roi, 1843), 41.

48. Auguste Laurent, "Action de l'acide sulfurique sur l'hydrure de benzoile," *Annales de chimie et de physique* 65 (1837): 192–204, 193.

49. Auguste Laurent, "Sur le benzoyle et la preparation de la benzimide; analyse de l'essence d'amandes amères," *Annales de chimie et de physique* [2] 60 (1835), 218.

50. Laurent, "Sur le benzoyle et la benzimide," *Annales de chimie et de physique* [2] 59 (1835), 397.

51. Laurent, "Sur le benzoyle et la benzimide," 403.

52. William E. Burns, "'A Proverb of Versatile Mutability': Proteus and Natural Knowledge in Early Modern Britain," *Sixteenth Century Journal* 32, no. 4 (Winter, 2001): 969–980, 972.

53. Laurent, "Action du chlore sur les hyrdochlorates d'ethérène et de methylene," *Annales de chimie et de physique* [2] 64 (1837), 334.

54. Marika Blondel-Mégrelis, "Auguste Laurent et les alcaloïdes," *Revue d'histoire de la pharmacie* 49, no. 331 (2001): 303–314, 311. He described it as hydrogen successively replacing the chlorine in what he called the "nucleus."

55. Laurent, "Sur la chlorophénise et les acides," 27.

56. Laurent, "Sur la chlorophénise et les acides," 44.

57. Laurent, "Sur la chlorophénise et les acides," "suitable space," 40, "no neighbors," 41.

58. J. Jacques, "La thèse de doctorat d'Auguste Laurent et la théorie des combinaisons organiques (1836)," *Bulletin de la Société Chimique de France* (1954): D31–39.

59. Satish Kapoor, "The Origins of Laurent's Organic Classification," *Isis* 60, no. 4 (1969): 477–527, 492–493.

60. "Annual Report, Anniversary Meeting of March 30, 1854," 155.

61. Leo Klosterman, "A Research School of Chemistry in the Nineteenth Century: Jean Baptiste Dumas and His Research Students," *Annals of Science* 42, no. 1 (1985): 41–80.

62. Jean Baptiste Dumas, *Leçons sur la philosophie chimique* (Paris: Ébrard, 1836), 4. See also Alan Rocke, *Chemical Atomism in the Nineteenth Century: From Dalton to Canizzarro* (Columbus: Ohio State University Press, 1984).

63. Dumas, *Leçons,* 223.

64. Dumas, *Leçons,* 383.

65. Laurent to Dumas, 12 June 1836, cited in J. Jacques, "Auguste Laurent et J.-B. Dumas d'après une correspondance inédite," *Revue d'histoire des sciences et de leurs applications* 6, no. 4 (1953): 329–349, 334.

66. Auguste Laurent, "Sur l'acide camphorique," *Annales de chimie et de physique* [2] 63 (1836): 215.

67. His thesis would ultimately have two components: *Recherches diverses de chimie organique et sur la densité des argiles mixtes,* and *Des considérations générales sur les propriétés physiques des atomes et sur leur forme.* Jacques, "La thèse de doctorat d'Auguste Laurent," D31–39.

68. Chaigneau, *Jean-Baptiste Dumas,* 136.

69. *Almanach du commerce de Paris* (Paris: Bureau de l'almanach du commerce, 1837), 351.

9. The Spirit of Coal Tar

1. Robert Stumper, "La vie et l'œuvre d'un grand chimiste, pionnier de la doctrine atomique: Augustin Laurent," *Archives de l'Institut grand-ducal de Luxembourg, section des sciences,* n.s. 20 (1953): 47–93, 53.

2. Dumas and Liebig, "Note sur l'état actuel de la chimie organique," *Comptes rendus des séances de l'Académie des Sciences* 5 (1837): 567–572, 569.

3. Dumas and Liebig, "Note sur l'état actuel de la chimie organique," 572.

4. J. Jacques, "La thèse de doctorat d'Auguste Laurent et la théorie des combinaisons organiques (1836)," *Bulletin de la Société Chimique de France* (1954): D31–39.

5. Marcel Chaigneau, *Jean-Baptiste Dumas: sa vie, son oeuvre, 1800–1884* (Paris: Guy Le Prat, 1984), 142.

6. Stumper, "La vie et l'œuvre d'un grand chimiste," 53.

7. Laurent to Dumas, cited in J. Jacques, "Auguste Laurent et J.-B. Dumas d'après une correspondance inédite," *Revue d'histoire des sciences et de leurs applications* 6, no. 4 (1953): 329–349, 335.

8. Laurent, "Recherches diverses de chimie organique," *Annales de chimie et de physique*[2] 66 (1837): 136–213 and 314–336.

9. Laurent to Dumas, cited in J. Jacques, "Auguste Laurent et J.-B. Dumas d'après une correspondance inédite," 337.

10. Leo Klosterman, "A Research School of Chemistry in the Nineteenth Century: Jean Baptiste Dumas and His Research Students," *Annals of Science* 42, no. 1 (1985): 41–80.

11. Alan J. Rocke, *Nationalizing Science: Adolphe Wurtz and the Battle for French Chemistry* (Cambridge, MA: MIT Press, 2001), 47.

12. Dumas to Liebig, n.d., postmarked 21 January 1838, Liebigiana IIB, cited in Rocke, *Nationalizing Science,* 111.

13. Pelouze to Liebig, 25 January 1838, Liebigiana IIB, cited in Rocke, *Nationalizing Science,* 108.

14. Wöhler to Liebig, 16 May 1832, cited in Catherine Jackson, "Analysis and Synthesis in Nineteenth-Century Organic Chemistry" (PhD diss., University of London, 2008), 45.

15. F. L. Holmes, "The Complementarity of Teaching and Research in Liebig's Laboratory," *Osiris* 5 (1989): 121–164, 147.

16. Christoph Meinel, "August Wilhelm Hofmann—'Reigning Chemist-in-Chief,'" *Angewandte Chemie* 31, no. 10 (October 1992): 1265–1398, 1266.

17. Dumas to Liebig, n.d., Liebigian IIB, cited in Rocke, *Nationalizing Science,* 109.

18. Rocke, *Nationalizing Science,* 113. Henry Freeman, "Rue Cuvier, rue Geoffroy-Saint-Hilaire, rue Lamarck: Politics and Science in the Streets of Paris," *Nineteenth-Century French Studies* 35, no. 3 / 4 (2007): 513–525.

19. Z. Delalande, "Recherches sur la coumarine ou stéaroptène des fèves de tonka," *Annales de chimie et de physique* [3] 6 (1841): 343–351.

20. Malaguti and Sarzeau, "Sur l'acide lithosélique," *Comptes rendus des séances de l'Académie des Sciences* 15 (1842): 518.

21. J.-S. Stas, "Recherches chimiques sur la Phlorizine, *Annales de chimie et de physique* [2] 69 (1838): 367–401. Charles Gerhardt, "Sur un nouveau mode de formation de acide valérianique," *Comptes rendus des séances de l'Académie des Sciences* 13 (1841): 309; Lewy, "Note sur la cire de Chine," *Comptes rendus des séances de l'Académie des Sciences* 17 (1843): 978; Scribe, "Note sur la matière amère du chardon-bénit," *Comptes rendus des séances de l'Académie des Sciences* 15 (1842): 802.

22. Edouard Grimaux, *Charles Gerhardt, sa vie, son œuvre, sa correspondance* (Paris: Masson, 1900), 32.

23. C. Gerhardt and A. Cahours, "Recherches chimiques sur les huiles essentielles," *Comptes rendus des séances de l'Académie des Sciences* 11 (1840): 900–902; Alexandre Étard, "Notice biographique sur Auguste Cahours," *Bulletin de la Société Chimique* 3rd series, 7 (1892): i—xii, i.

24. Mi Gyung Kim, "Constructing Symbolic Spaces: Chemical Molecules in the Académie des Sciences," *Ambix* 43 (1996): 1–31.

25. Liebig to Berzelius, 22 July 1834, cited in W. H. Brock, *Justus von Liebig: The Chemical Gatekeeper* (Cambridge: Cambridge University Press, 1997), 72.

26. Brock, *Justus von Liebig*, 85.

27. Marika Blondel-Mégrelis, "Liebig or How to Popularize Chemistry," *Hyle* 13, no. 1 (2007): 29–40, 15.

28. S. C. H. Windler, "Ueber das Substitutionsgesetz und die Theorie der Typen," *Annalen der Chemie und Pharmacie* 33 (1840): 308–310.

29. Windler, "Ueber das Substitutionsgesetz," 310.

30. Jules Mersch, "Mathieu-Lambert Schrobilgen, 1789–1883," *Biographie nationale du pays de Luxembourg*, fasc. I (Luxembourg: Imprimerie de la Cour Victor Buck, 1947), 63.

31. Eugen Weber, *A Modern History of Europe* (1971; London: Hale, 1973), 448.

32. Stumper, "La vie et l'œuvre d'un grand chimiste," 56.

33. Laurent to Gerhardt, 12 February 1845, *Correspondance de Charles Gerhardt*, vol. 1: *Laurent et Gerhardt; lettres échangées entre Auguste Laurent et Charles Gerhardt, 1844–1852*, ed. Marc Tiffeneau (Paris: Masson, 1918), 18.

34. Laurent to Gerhardt, 12 February 1845, cited in *Correspondance de Charles Gerhardt*, 19.

35. Charles Gerhardt first proposed the name phenol. Charles Gerhardt, "Recherches sur la salicine," *Annales de chimie et de physique*, [3] 7 (1843): 215–229, 221.

36. Auguste Laurent, "Mémoire sur le phényle et ses dérivés," *Annales de chimie et de physique* [3] 3 (1841): 195–228, 198.

37. Laurent, "Mémoire sur le phényle et ses dérivés," 198.

38. Auguste Laurent, "Mémoire sur la série stilbique," *Comptes rendus des séances de l'Académie des Sciences* 16 (1843): 856–860, 856.

39. Laurent, "Mémoire sur la série stilbique," 857.

40. Catherine E. McKinley, *Indigo: In Search of the Colour That Seduced the World* (London: Bloomsbury, 2011); Andrea Feeser, *Red, White, and Black Make Blue: Indigo in the Fabric of Colonial South Carolina Life* (Athens: University of Georgia Press, 2013).

41. Partington, *A History of Chemistry*, vol. 4 (New York: Macmillan, 1964), 389; Laurent, "Mémoire sur le phényle et ses dérivés."

42. J. Dumas, "Quatrième mémoire sur les types chimiques," *Annales de chimie et de physique* [3] 2 (1841): 204–232, 223.

43. Auguste Laurent, "Trente et unieme memoire sur les types ou radicaux dérivés (qui n'ont pas été inventés par M. Dumas)," *Revue scientifique de Quesneville* (1842) 9: 5–34.

44. Laurent, "Mémoire sur le phényle et ses dérivés."

45. Jöns Jacob Berzelius, *Rapport annuel sur les progrès des sciences physiques et chimiques, présenté le 31 mars 1842 à l'Académie royale des sciences de Stockholm* (Paris: Fortin, Masson, et Cie, 1843), 256.

46. Berzelius, *Rapport annuel*, 193.

47. Kim, "Constructing Symbolic Spaces," 27.

48. Gerhardt to Liebig, Montpellier, 8 December 1842, in Grimaux and Gerhardt, *Charles Gerhardt*, 459.

49. Liebig to Gerhardt, 1 March 1840, in Grimaux, *Charles Gerhardt*, 42, 43.

50. Laurent to Gerhardt, 5 January 1846, *Correspondance de Charles Gerhardt*, n. 28, 121.

51. Mersch, "Mathieu-Lambert Schrobilgen," 48.

52. Laurent to Gerhardt, 5 January 1846.

53. August Hofmann recalled the event in his President's Address, although he placed it in 1844. Much of the secondary literature follows him on this date, although it must be mistaken as their joint work together was published in 1843. "Annual Report, Read by the President Philip Yorke, Anniversary Meeting of March 30, 1854" ("President's Address"), *Quarterly Journal of the Chemical Society of London* 7, no. 26 (1855): 144–159, 153.

54. Brock, *Justus von Liebig*, 60.

55. Cited in Meinel, "August Wilhelm Hofmann," 1266.

56. Meinel, "August Wilhelm Hofmann," 1266.

57. August Hofmann, "Chemische Untersuchung der organischen Basen im Steinkohlen-Theeroel" (Chemical investigation of the organic bases in coal-tar oil), *Annalen der chemie und pharmacie* 47 (1843): 37–87.

58. Laurent, "Sur un nouveau mode de formation de l'aniline," *Comptes rendus des séances de l'Académie des Sciences* 17 (1843): 1366–1368, 1368.

59. Laurent, "Sur un nouvel alkali organique, l'amarine," *Comptes rendus des séances de l'Académie des Sciences* 19 (1844): 353–355, 353.

60. McKinley, *Indigo*, 2; Prakash Kumar, *Indigo Plantations and Science in Colonial India* (Cambridge: Cambridge University Press, 2012).

61. Laurent, "Sur un nouveau mode de formation de l'aniline," 1368.

62. Mersch, "Mathieu-Lambert Schrobilgen," 48.

63. Auguste Laurent, "Sur la composition des alcalis organiques et de quelques combinaisons azotées," *Annales de chimie et de physique* 19 (1847): 359–377.

64. Laurent to Gerhardt, 6 March 1845, *Correspondance de Charles Gerhardt*, 28.

65. J. Persoz, "Note sur les acides amidés et sur la constitution moléculaire de divers composés organiques," *Compte rendus de l'Académie des Sciences* 19 (1844): 435–439, 436.

66. Laurent to Gerhardt, 4 July 1844, *Correspondance de Charles Gerhardt*, 4.

67. Laurent, "Sur un nouvel alkali organique, l'amarine."

68. Quesneville, *Revue scientifique et industrielle*, vol. 4 (Paris: Hachette, 1841).

69. Dumas, "Rapport sur un Mémoire de M. Cahours relative à l'huile volatile de Gaultheria procumbens," *Comptes rendus des séances de l'Académie des Sciences* 18 (1844): 287–289, 287.

70. Apollinaire Bouchardat, "Sur les propriétés optiques des alcalis végétaux," *Annales de chimie et de physique* [3] 9 (1843): 213–244, "active substance," 226; "marked and diverse," 244.

71. Bouchardat, "Sur les propriétés optiques des alcalis végétaux," 241.

72. Jean-Baptiste Biot, "Note sur un travail de M. Bouchardat, relative aux alcalis végétaux," *Comptes rendus des séances de l'Académie des Sciences* 17 (1843): 721–724.

73. Auguste Laurent, "Action de quelques bases organiques sur la lumière polarisée," *Comptes rendus des séances de l'Académie des Sciences* 19 (1844): 925.

74. Laurent, "Action des alcalis chlorés sur la lumière polarisée et sur l'économie animale," *Comptes rendus des séances de l'Académie des Sciences* 19 (1844): 219.

75. Laurent, "Action de quelques bases organiques sur la lumière polarisée," 925.

10. THE STUDY OF THINGS THAT DO NOT EXIST

1. Laurent to Gerhardt, 29 October 1845, *Correspondance de Charles Gerhardt*, vol. 1: *Laurent et Gerhardt; lettres échangées entre Auguste Laurent et Charles Gerhardt, 1844–1852*, ed. Marc Tiffeneau (Paris: Masson, 1918), 111.

2. Christoph Meinel, "August Wilhelm Hofmann—'Reigning Chemist-in-Chief,'" *Angewandte Chemie* 31, no. 10 (1992): 1265–1398, 1268.

3. John J. Beer, "A. W. Hofmann and the Founding of the Royal College of Chemistry," *Journal of Chemical Education* 37 (1960): 248–251, 248.

4. Meinel, "August Wilhelm Hofmann," 1268.

5. Meinel, "August Wilhelm Hofmann," 1269. See also Catherine M. Jackson, "Re-Examining the Research School: August Wilhelm Hofmann and the Re-Creation of a Liebigian Research School in London," *History of Science* 44, no. 3 (2006): 281–319.

6. Charles Kingsley, "Memoir," in C. B. Mansfield, *Paraguay, Brazil, and the Plate* (Cambridge: Macmillan, 1856), xii.

7. *Reports of the Royal College of Chemistry* (London: Schulze and Co., 1849), lxiii.

8. C. B. Mansfield, "Fabrication et rectification des essences et huiles essentielles du goudron de houille," *Le Technologiste* 10 (October 1848): 13–22.

9. "Exposition Universelle de Londres," *Cosmos: Revue encyclopédique hebdomadaire des progrès des sciences* 21 (1862): 512.

10. *Les Mondes*, vol. 14 (Paris: Librairie des Mondes, 1867), 154.

11. Henk Van Den Belt, "Why Monopoly Failed: The Rise and Fall of Société La Fuchsine," *British Journal for the History of Science* 25, no. 1 (1992): 45–63.

12. Charles Girard and Georges de Laire, *Traité des dérivés de la houille* (Paris: G. Masson, 1873), 252.

13. Charles Blachford Mansfield, "Manufacture and Application of Products from Coal: Benzine," *Journal of the Society of Arts* 3 (1855): 104–105, 105.

14. "Exposition Universelle de Londres," *Cosmos*, 513.

15. William H. Brock, *The Chemical Tree: A History of Chemistry* (New York: Norton, 2000), 221.

16. Quoted in Josette Fournier, "Auguste Laurent (1807–1853) dans la revue scientifique du Dr Quesneville," *Revue d'histoire de la pharmacie* 359 (2008): 287–303.

17. Edouard Grimaux, *Charles Gerhardt, sa vie, son œuvre, sa correspondance* (Paris: Masson, 1900), 203.

18. Grimaux, *Charles Gerhardt*, 96.

19. Laurent to Gerhardt, 11 May 1845, no. 10, *Correspondance de Charles Gerhardt*, 43.

20. Laurent to Gerhardt, 14 February 1845, no. 4, *Correspondance de Charles Gerhardt*, 19.

21. Laurent and Gerhardt, "Sur les dérivées de la morphine et de la narcotine," *Annales de chimie et de physique*, 24 (September 1845): 112; Laurent and Gerhardt, "Sur de nouvelles combinaisons de l'essence d'amandes amères," *Annales de chimie et de physique* 30 (1850): 404.

22. Auguste Laurent, "Sur de nouvelles combinaisons chlorurées de la naphtaline, et sur l'isomorphisme et l'isomérie de cette série," *Compte rendus de l'Académie des Sciences* 16 (1842): 818–822, 819.

23. Auguste Laurent, "Note sur les combinaisons organiques," *Compte rendus de l'Académie des Sciences* 17 (1843) : 311–312, 312.

24. Cited in Joseph Fruton, *Methods and Styles in the Development of Chemistry* (Philadelphia: American Philosophical Society, 2002), 87–88.

25. Laurent to Gerhardt, December 1844, no. 2, *Correspondance de Charles Gerhardt,* 11.

26. Laurent to Gerhardt, 19 April 1845, no. 9, *Correspondance de Charles Gerhardt,* 37.

27. Laurent to Gerhardt, 29 October 1845, no. 25,*Correspondance de Charles Gerhardt,*110.

28. Gerhardt to Laurent, 17 May 1845, cited in Marika Blondel-Mégrelis, "Quelques aspects méconnus de la personne et de l'oeuvre de Charles Gerhardt (1816–1856)," *Revue d'histoire de la pharmacie* 357 (2008): 39–62, 59.

29. Laurent to Gerhardt, 9 April 1845, no. 8, *Correspondance de Charles Gerhardt,* 35.

30. Laurent to Gerhardt, 28 June 1845, no. 17, *Correspondance de Charles Gerhardt,* 75.

31. Clara deMilt, "Auguste Laurent, Founder of Modern Organic Chemistry," *Chymia* 4 (1953): 85–114.

32. Laurent to Gerhardt, 14 September 1845, no. 19, *Correspondance de Charles Gerhardt,* 82.

33. Quoted in Robert Stumper, "La vie et l'œuvre d'un grand chimiste, pionnier de la doctrine atomique: Augustin Laurent," *Archives de l'Institut grand-ducal de Luxembourg, section des sciences,* n.s. 20 (1953): 47–93, 62.

34. Laurent to Gerhardt, 14 September 1845, no. 19, *Correspondance de Charles Gerhardt,* 83.

35. Laurent to Gerhardt, 14 September 1845, no. 23, *Correspondance de Charles Gerhardt,* 101.

36. Laurent, "Sur le mode de combinaison des corps," *Compte rendus de l'Académie des Sciences* 21 (1845), 852–860, 860, quoted in John Hedley Brooke, "Laurent, Gerhardt, and the Philosophy of Chemistry," *Historical Studies in the Physical Sciences* 6 (1975): 405–429.

37. Laurent, "Sur le mode de combinaison des corps," 853.

38. Laurent, "Sur le mode de combinaison des corps," 860.

39. Laurent to Gerhardt, 29 October 1845, no. 25, *Correspondance de Charles Gerhardt,* 107, italics in original.

40. Laurent to Gerhardt, Luxembourg, 29 October 1845, no. 25, *Correspondance de Charles Gerhardt,* 108.

41. Patrice Debré, *Louis Pasteur,* trans. Elborg Forster (Baltimore: Johns Hopkins University Press, 1998), 30.

42. Gerald Geison, *The Private Science of Louis Pasteur* (Princeton: Princeton University Press, 1995), 61.

43. René Vallery-Radot, *The Life of Pasteur,* trans. R. L. Devonshire, 2 vols. (London: Constable, 1911), 1: 32.

44. Seymour Mauskopf, "Crystals and Compounds," *Transactions of the American Philosophical Society,* n.s. 66, part 3 (1976): 1–82, 71.

45. Stumper, "La vie et l'œuvre d'un grand chimiste," 65.

46. Grimaux, *Charles Gerhardt,* 206.

47. Laurent to Gerhardt, February 23 1847, no. 65, *Correspondance de Charles Gerhardt,* 227.

48. DeMilt, "Auguste Laurent," 106.

49. Laurent to Gerhardt, 30 May 1847, no. 68, *Correspondance de Charles Gerhardt,* 239.

50. Laurent to Gerhardt, 30 May 1847, no. 68, *Correspondance de Charles Gerhardt,* 241.

51. Vallery-Radot, *Life of Pasteur,* 34.

52. Laurent to Gerhardt, Paris, 28 October 1847, no. 71, *Correspondance de Charles Gerhardt,* 247.

53. Laurent to Gerhardt, Paris, 6 October 1847, no. 70, *Correspondance de Charles Gerhardt,* 245.

54. *The Fifth Exhibition of the Massachusetts Charitable Mechanic Association* (Boston: Dutton and Wentworth, 1848), 171.

55. A. McElroy, *Philadelphia Directory 1839* (Philadelphia: Isaac Ashmead & Co., 1839), unpaged.

56. *Gazette des Tribunaux*, no. 4304, 27 June 1839; *Gazette des Tribunaux*, no. 4456, 20 December 1839.

57. *Revue scientifique et industrielle* 6 (1845).

58. The Fourth Exhibition of the Massachusetts Charitable Mechanic Association (Boston: Crocker and Brewster, 1844), 79.

59. "The First Syruped Soda Water in America," *Pharmaceutical Era*, February 1913, 64. Tristan Donovan, *Fizz: How Soda Shook Up the World* (Chicago: Chicago Review Press, 2014), 35; Tara Dixon, "Scents and Soda: French Perfume, American Glassworks, and the Rise of the Retail Water Industry," *Pennsylvania Magazine of History and Biography* 142, no. 3 (2018): 239–267.

60. Eugène Roussel, "Notes et Recettes de Liquers de Tables," n.d., Eugène Roussel Papers, collection no. 1785, E-90, box 2.

61. École supérieure de pharmacie de Paris, *Codex pharmacopée française* (Paris: Béchet jeune, 1837), 187.

62. *Philadelphia Public Ledger*, 10 July 1839.

63. *Philadelphia Public Ledger*, 26 June 1855.

64. *Philadelphia Public Ledger*, 7 April 1855.

65. Weather Journal, 1855, entry for June 28, Eugène Roussel Papers, box 2, collection no. 1785, Historical Society of Pennsylvania, Philadelphia.

66. *Philadelphia Public Ledger*, 17 May 1855.

67. *Evening Argus*, 27 June 1855.

11. THE SYNTHETIC AGE

1. *Revue scientifique et industrielle* 6 (1845); *Annuaire général du commerce, de l'industrie, de la magistrature et de l'administration* (Paris: Firmon Didot, 1847), 170.

2. Malcolm Crook, *How the French Learned to Vote: A History of Electoral Practice in France* (Oxford: Oxford University Press, 2021), 159.

3. Jean Reynaud, *Revue encyclopédique*, January 1832, cited in David A. Griffiths, "Jean Reynaud: An Unfamiliar Page from the History of Socialist Thought," *Science and Society* 46, no. 3 (1982): 361–368, 362.

4. Victor Hugo, *Histoire d'un crime: Cahier complementaire, Oeuvres complètes de Victor Hugo* (Paris: l'Imprimerie Nationale, 1907), 241.

5. Jill Harsin, *Barricades: The War of the Streets in Revolutionary Paris, 1830–1848* (New York: Palgrave Macmillan, 2002), 122, 125n2.

6. Charles Robin, *Histoire de la révolution française de 1848* (Paris: Penaud, 1849), 206.

7. Victor Hugo, *History of a Crime,* trans. Huntington Smith, 2 vols. (Boston: T. Y. Crowell, 1888), 2: 48; Victor Hugo, *Histoire d'un crime: Cahier complementaire, Œuvres complètes de Victor Hugo* (Paris: l'Imprimerie Nationale, 1907), 33.

8. Hugo, *Histoire d'un crime,* 34. Hugo, *History of a Crime,* 2: 48.

9. Percy Bolingbroke St. John, *French Revolution in 1848: The Three Days of February, 1848* (London: Richard Bentley, 1848), 94.

10. Mike Rapport, *1848: Year of Revolution* (New York: Basic Books, 2009), 51.

11. St. John, *French Revolution in 1848,* 163.

12. "Jean Reynaud: French Mystic and Philosopher," *Fraser's Magazine for Town and Country* 17 (June 1878): 718–728.

13. Nicklès, "Correspondence of M. Jerome Nicklès, dated April 27, 1853," *American Journal of Science and the Arts* 16 [2nd series] (November 1853): 103–104, 104; *Correspondance de Charles Gerhardt,* vol. 1: *Laurent et Gerhardt; lettres échangées entre Auguste Laurent et Charles Gerhardt, 1844–1852,* ed. Marc Tiffeneau (Paris: Masson, 1918), 265, 292. *Revue de Paris* 24 (1855): 302.

14. René Vallery-Radot, *The Life of Pasteur,* trans. R. L. Devonshire, 2 vols. (London: Constable, 1911), 1: 37.

15. Vallery-Radot, *Life of Pasteur,* 37.

16. Gerald Geison, *The Private Science of Louis Pasteur* (Princeton: Princeton University Press, 1995), 53.

17. Patrice Debré, *Louis Pasteur,* trans. Elborg Forster (Baltimore: Johns Hopkins University Press, 1998), 7.

18. Vallery-Radot, *Life of Pasteur,* 15.

19. Debré, *Louis Pasteur,* "baths," 22, 26.

20. Jean-Joseph Pasteur to his son, Arbois, 11 December 1846, *Correspondance de Pasteur 1840–1895,* ed. Louis Pasteur Vallery-Radot, 4 vols. (Paris: B. Grasset, 1940), 1: 147.

21. Pasteur to his parents, 9 December 1842, *Correspondance de Pasteur, 1840–1895*, 1: 81, cited in Leo Klosterman, "A Research School of Chemistry in the Nineteenth Century: Jean Baptiste Dumas and His Research Students: Part 1," *Annals of Science* 42, no. 1 (1985): 1–40, 6.

22. Geison, *The Private Science of Louis Pasteur*, 60.

23. Geison, *The Private Science of Louis Pasteur*, 61; Seymour Mauskopf, "Crystals and Compounds," *Transactions of the American Philosophical Society*, n.s. 66, part 3 (1976): 1–82.

24. Aaron J. Ihde, *The Development of Modern Chemistry* (New York: Harper and Row, 1964), 172; Marc Drouot, André Rohmer, and Nicolas Stoskopf, *La fabrique de produits chimiques Thann et Mulhouse: histoire d'une entreprise de 1808 à nos jours* (Strasbourg: La Nuée bleue, 1991).

25. Biot, "Mémoire sur une expérience de M. Mitscherlich, concernant les caractères optiques du paratrarte et du tartrate double de soude et d'ammoniaque," *Compte rendus de l'Académie des Sciences* 19 (1844): 721.

26. Biot, "Mémoire sur une expérience de M. Mitscherlich," 719.

27. Pasteur, "La Dissymétrie moléculaire," *Oeuvres de Pasteur*, vol. 1 (Paris: Masson et Cie, 1922), 371.

28. Pasteur, "La Dissymétrie moléculaire," 371.

29. Geison, *The Private Science of Louis Pasteur*, 84.

30. Vallery-Radot, *The Life of Pasteur*, 39.

31. Vallery-Radot, *The Life of Pasteur*, 46.

32. Vallery-Radot, *The Life of Pasteur*, 40, 41.

33. Vallery-Radot, *The Life of Pasteur*, 54.

34. Christoph Meinel, "August Wilhelm Hofmann—'Reigning Chemist-in-Chief,'" *Angewandte Chemie* 31, no. 10 (1992): 1265–1398, 1271; Hofmann to the Ministry (1873), Geheimes Staatsarchiv (Merseburg), Rep. 76, Va, Sec. 2, Tit. XV, no. 70, vol. 2, 67.

35. Chevreul was also not able to make it, but he did send his excuses. Edouard Grimaux and Charles Gerhardt, *Charles Gerhardt, sa vie, son oeuvre, sa correspondance 1816–1856* (Paris: Masson et Cie, 1900), 222, cited in John Buckingham, *Chasing the Molecule* (Stroud: Sutton, 2005), 124.

36. Catherine M. Jackson, "Synthetical Experiments and Alkaloid Analogues: Liebig, Hofmann, and the Origins of Organic Synthesis," *Historical Studies in the Natural Sciences* 44, no. 4 (2014): 319–363. More generally, see Bernadette Bensaude-Vincent and William R. Newman, eds., *The Artificial and the Natural: An Evolving Polarity* (Cambridge, MA: MIT Press, 2007).

37. J. S. Muspratt and A. W. Hofmann, "Über das Toluidin, eine neue organische Basis," *Annalen der Chemie und Pharmacie* 54 (1855): 1–29, 3, translation by Meinel, "August Wilhelm Hofmann," 1270.

38. Paul Hollister, "The Glazing of the Crystal Palace," *Journal of Glass Studies* 16 (1974): 95–110, 106.

39. Charles Dodgson to his sister Elizabeth, 5 July 1851, *The Selected Letters of Lewis Carroll*, ed. Roger Lancelyn Green (London: Palgrave MacMillan, 1989), 12.

40. Catherine Maxwell, *Scents and Sensibility: Perfume in Victorian Literary Culture* (Oxford: Oxford University Press, 2017), 22.

41. Hofmann, "Application of Organic Chemistry to Perfumery. From a Letter Written by Dr. Hofmann to Prof. Liebig," *Chemical Gazette* 10 (1852): 98–100, "pear," 98; "doubtful," 100.

42. *Reports by the Juries: On the Subjects in the Thirty Classes into Which the Exhibition Was Divided*, vol. 1 (London: William Clowes & Sons, 1852), 609.

43. *Reports by the Juries*, 1: 605.

44. "Hoffman's [sic] letter to Liebig," *Annual of Scientific Discovery*, David A. Wells, ed. (Boston: Gould and Lincoln, 1853) : 227.

45. Collas published his secret in 1851 in Gustave-Augustin Quesneville, *Secrets des arts* (Paris, 1851), 4: 215.

46. Elisabeth Celnart, *Nouveau manuel complet du parfumeur* (Paris: Roret, 1854), 246.

47. David Barnes, *The Making of a Social Disease: Tuberculosis in Nineteenth-Century France* (Berkeley: University of California Press, 1995).

48. Marcel Chaigneau, *Jean-Baptiste Dumas: sa vie, son oeuvre, 1800–1884* (Paris: Guy Le Prat, 1984), 210. Alan J. Rocke, *Nationalizing Science: Adolphe Wurtz and the Battle for French Chemistry* (Cambridge, MA: MIT Press, 2001), 142.

49. Vallery-Radot, *The Life of Pasteur*, 56.

50. Vallery-Radot, *The Life of Pasteur,* 56.

51. Marika Blondel-Mégrelis, "Quelques aspects méconnus de la personne et de l'œuvre de Charles Gerhardt (1816–1856)," *Revue d'histoire de la pharmacie* 357 (2008): 39–62, 52.

52. Geison, *The Private Science of Louis Pasteur,* 86. Geison counts Laurent's name appearing thirteen times in Pasteur's writings before his discovery in 1848, and only once in the entirety of his subsequent life.

53. Pasteur, *Correspondance de Pasteur 1840–1895,* 1: 236, cited in Geison, *The Private Science of Louis Pasteur,* 88.

54. Laurent to Gerhardt, 13 June 1851, no. 77, *Correspondance de Charles Gerhardt,* 267.

55. " "In seeing all these trees and flowers, all these plants coming out of the ground, I set myself to cursing the Biots and Chaptals." Laurent to Biot, 25 April 1851, Bibliothèque de l'Institut, Dossier Biot.

56. Laurent to Dumas, 14 December 1851, no. 4. J. Jacques, "Auguste Laurent et J.-B. Dumas d'après une correspondance inédite," *Revue d'histoire des sciences et de leurs applications* 6, no. 4 (1953): 329–349, 340.

57. *Le Tcheou-Li ou Rites des Tcheou,* trans. Édouard Biot, ed. Jean-Baptiste Biot (Paris: l'Imprimerie Nationale, 1851).

58. Grimaux, *Charles Gerhardt,* 242.

59. Chaigneau, *Jean-Baptiste Dumas,* 147.

60. Rocke, *Nationalizing Science,* 140; Flourens, Perpetual Secretary of the Academy of Sciences, to Madame Laurent, 8 December 1853, Archives of the Bibliothèque de l'Institut, Fonds Pellier-Laurent, #201.

61. Letter of Jean-Baptiste Dumas, n.d., Archives of the Academy of Sciences, Dossier Laurent.

62. Cited in Patrice Higonnet, *Paris: Capital of the World* (Cambridge, MA: Belknap Press of Harvard University Press, 2005), 10. See also David Harvey, *Paris: Capital of Modernity* (New York: Routledge, 2003).

63. Priscilla Parkhurst Ferguson, *Paris as Revolution: Writing the Nineteenth-century City* (Berkeley: University of California Press, 1997), 119.

64. Higonnet, *Paris,* 175; see also Louis Veuillot, *Odeurs de Paris* (Paris: Palmé, 1867); David Barnes, *The Great Stink of Paris and the Nineteenth-Century Struggle against Filth and Germs* (Baltimore: Johns Hopkins University Press, 2006).

65. Colin Jones, *Paris: Biography of a City* (New York: Penguin, 2006), 318.

66. Haussmann, *Mémoires,* vol. 3 (Paris: Victor-Havard, 1893), 825, cited in Jones, *Paris,* 318.

67. Karl Marx, *The Eighteenth Brumaire of Louis Napoleon,* trans. Saul K. Padover and Friedrich Engels (Moscow: Progress Publishers, 1937), 1.

68. Ernest Alfred Vizetelly, *The Court of the Tuileries, 1852–1870* (London: Chatto and Windus, 1907), 9. David P. Jordan, *Transforming Paris: The Life and Labors of Baron Haussman* (New York: Free Press, 1995).

69. Rufus Frost-Herrick, *Denatured or Industrial Alcohol* (New York: Wiley and Sons, 1907), 3.

70. Paris Archives, État Civil, 1860–1872 Décès, 10, D1M9 761.

12. LIFE IS ASYMMETRIC

1. August Hofmann, "On Insolinic Acid," *Proceedings of the Royal Society* 8 (1855): 1–3.

2. Edward Ward, "The Death of Charles Blachford Mansfield (1819–1855)," *Ambix* 31, no. 2 (1984): 68–69, 68.

3. Simon Garfield, *Mauve: How One Man Invented a Color That Changed the World* (New York: Norton, 2001), 19. Anthony S. Travis, "Perkin's Mauve: Ancestor of the Organic Chemical Industry," *Technology and Culture* 31, no. 1 (1990): 51–82.

4. William Hudson Brock, ed., *Justus von Liebig und August Wilhelm Hofmann in ihren Briefen (1841–73)* (Weinheim: Verlag Chemie, 1984), 14–17.

5. Garfield, *Mauve,* 77.

6. Quoted in Garfield, *Mauve,* 57.

7. W. H. Perkin, "On the Artificial Production of Coumarin and Formation of Its Homologues," *Journal of the Chemical Society* 21 (1868): 53–63, 55.

8. Garfield, *Mauve,* 4.

9. Otto Witt, "Obituary Notice, Rudolph Fittig, 1835–1910," *Journal of the Chemical Society, Transactions* 99 (1911).

10. Anthony Travis, "Science's Powerful Companion: A. W. Hofmann's Investigation of Aniline Red and Its Derivatives," *British Journal for the History of Science* 25, no. 1 (1992): 27–44, 35.

11. Eugénie Briot, "Les Fabriques de Laire, pionnières de la chimie des corps odorantes à Calais (1876–1914)," *PME et grandes entreprises en Europe du Nord-Ouest XIXe–XXe siècle: activités, strategies, performances* (Villeneuve d'Ascq: Presses Universitaires du Septentrion, 2012), 137–148.

12. Eugénie Briot, "From Industry to Luxury: French Perfume in the Nineteenth Century," *Business History Review* 85, no. 2 (2011): 273–294, 282.

13. Robert Bienaimé, "Une grande figure de la parfumerie française: Paul Parquet," *Industrie de la Parfumerie* (October 1955): 409–411.

14. Patricia de Nicola, "A Smelling Trip into the Past: The Influence of Synthetic Materials on the History of Perfumery," *Chemistry & Biodiversity* 5, no. 6 (2008): 1137–1148, 1138; Michael Edwards, *Perfume Legends, French Feminine Fragrances* (Paris: HM Editions, Levallois, 1996).

15. Briot, "From Industry to Luxury," 294. For more on the rise of modern perfumery, see Richard Stamelman, *Perfume: Joy, Obsession, Scandal, Sin: A Cultural History of Fragrance from 1750 to the Present* (New York: Rizzoli, 2006); Edwin T. Morris, *Fragrance: The Story of Perfume from Cleopatra to Chanel* (New York: Charles Scribner, 1984); Rosine Lheureux, *Une histoire des parfumeurs: France, 1850–1910* (Ceyzérieu: Champ Vallon, 2016); Jean-Louis Kupper, "En marge de la revolution industrielle: La grande parfumerie aux XIXe et XXe siècles," *Bulletin de la classe des lettres et des sciences morales et politiques* 16, no. 1 / 6 (2005): 11–35.

16. "Beautiful Tar," *Punch, or the London Charivari*, September 15, 1888, 123.

17. A. T. Demeyer, "L'aniline et les couleurs d'aniline," *Journal de chimie médicale, de pharmacie et de toxicology*, vol. 2, 5th series (1866): 99–102, 101.

18. R. Prosser White, "The Toxic Effects of Nitrobenzol," *The Practitioner* 43 (July to December, 1889): 14.

19. Jüdell, *Die Vergiftungen mit Blausäure u. Nitrobenzol in forensischer Beziehung* (Erlangen, 1876).

20. Gerald Geison, *The Private Science of Louis Pasteur* (Princeton: Princeton University Press, 1995), 106.

21. Louis Pasteur, “Mémoire sur la fermentation alcoolique,” *Annales de chimie et de physiques* 58 (1860): 323–426, 359.

22. Geison, *The Private Science of Louis Pasteur,* 101.

23. Pasteur to Chappuis, cited in René Vallery-Radot, *The Life of Pasteur,* trans. R. L. Devonshire, 2 vols. (London: Constable, 1911), 1: 113.

24. Pasteur, *Correspondance de Pasteur 1840–1895,* ed. Louis Pasteur Vallery-Radot, 4 vols. (Paris: B. Grasset, 1940), 1: 227, cited in Geison, *The Private Science of Louis Pasteur,* 138.

25. Geison, *The Private Science of Louis Pasteur,* 138.

26. Louis Pasteur, “La dissymétrie moléculaire, conférence faite à la société chimique de Paris le 22 décembre 1883,” *Oeuvres de Pasteur,* vol. 1: *Dissymétrie moléculaire* (Paris: Masson et Cie, 1922), 376.

27. Pasteur, “La dissymétrie moléculaire, conférence faite à la société chimique,” 376.

28. Geison, *The Private Science of Louis Pasteur,* 138.

29. Geison, *The Private Science of Louis Pasteur,* 104. For the Chemical Society of Paris, see Marika Blondel-Mégrelis, “Esquisse pour une histoire de la Société chimique, 1857–2007,” *L'Actualité chimique* 310 (July 2007): i–xx; Ulrike Fell and Alan Rocke, “The Chemical Society of France in Its Formative Years, 1857–1914,” in *Creating Networks in Chemistry: The Foundation and Early History of Chemical Societies in Europe,* ed. Anita Kildebaek Nielson and Sona Strbánová, 91–112 (London: Royal Society of Chemistry, 2008).

30. Pasteur, “La dissymétrie moléculaire, conférence faite à la société chimique,” 373.

31. Geison, *The Private Science of Louis Pasteur,* 135. Geison says he was committed to this doctrine since 1851, probably from his first work with Biot and Laurent.

32. Geison, *The Private Science of Louis Pasteur,* 105.

33. Pasteur, “Réponses aux remarques de MM. Wyrouboff et Jungfleisch sur La dissymétrie moléculaire,” *Oeuvres de Pasteur,* vol. 1: *Dissymétrie moléculaire,* 386.

34. Vallery-Radot, *The Life of Pasteur,* 114.

35. Vallery-Radot, *The Life of Pasteur,* 114.

36. Vallery-Radot, *The Life of Pasteur,* 115.

37. Vallery-Radot, *The Life of Pasteur,* 115.

38. *La Presse,* 1860, cited in Vallery-Radot, *The Life of Pasteur,* 129.

39. Vallery-Radot, *The Life of Pasteur,* 100.

40. Geison, *The Private Science of Louis Pasteur,* 110.

41. Leeuwenhoek to H. Oldenburg, 9 October 1676, incomplete English translation published in *Philosophical Transactions* 12, no. 133 (1677): 821–831. (The phrase "living atom" is not in the original but was added to the *Philosophical Transactions* text.)

42. Biot, "Avis au lecteur," in Auguste Laurent, *Méthode de chimie* (publié posthume par Jean-Baptiste Biot) (Paris: Imprimerie de Mallet Bachelier, 1854), vi.

43. Laurent, *Méthode de chimie,* xv, "complete polyhedron," 408.

44. Rocke, *The Quiet Revolution: Hermann Kolbe and the Science of Organic Chemistry* (Berkeley: University of California Press, 1993), "belong," 164, "seduced," 161.

45. Rocke, *Image and Reality: Kekulé, Kopp, and the Scientific Imagination* (Chicago: University of Chicago Press, 2010), 42.

46. Quoted in Rocke, *Image and Reality,* 194.

47. For the role of Laurent's Method of Chemistry, see Joseph Wotiz and Susanna Rudofsky, "Kekule's Dreams: Fact or Fiction," *Chemistry in Britain* 20 (1984): 720; John Wotiz, ed., *The Kekulé Riddle: A Challenge for Chemists and Psychologists* (Carbondale, IL: Cache River Press, 1993), 221–245. Alan Rocke has shown that Laurent was not in fact proposing that benzene was made up of a hexagon of linked carbon atoms, but rather called it a "polyèdre complet." O. B. Ramsay and Alan Rocke, "Kekulé's Dreams: Separating the Fiction from the Fact," *Chemistry in Britain* 20 (1984): 1093–1094; Alan Rocke, "It Began with a Daydream: The 150th Anniversary of the Kekulé Benzene Structure," *Angewandte Chemie* 54, no. 1 (2015): 46–50.

48. Christoph Meinel, "Molecules and Croquet Balls," in *Models: The Third Dimension of Science,* ed. S. De Chadarevian and N. Hopwood (Stanford: Stanford University Press, 2004), 259.

49. In fact, van 't Hoff published several pamphlets. The first, published 5 September 1874, was in Dutch and can be translated as *A Proposal for Extending the Currently Employed Structural Formulae in chemistry into Space, Together With a Related Remark on the Relationship between Optical Activating Power and Chemical Constitution of Organic Compounds.* He published a French version in 1875 called *La chimie dans l'espace* and a German version in 1877 called *Die Lagerung der Atome im Raume.* See Peter Ramberg and Geert Somsen, "The Young J. H. van 't Hoff: The Background to the Publication of His 1874 Pamphlet on the Tetrahedral Carbon Atom, Together with a New English Translation," *Annals of Science* 58 (2001): 51–74; O. Bertrand Ramsay, ed., *Van 't Hoff-Le Bel Centennial* (Washington, DC: American Chemical Society, 1975); Peter Ramberg, *Chemical Structure, Spatial Arrangement: The Early History of Stereochemistry, 1874–1914* (New York: Routledge, 2017).

50. Ramberg, *Chemical Structure, Spatial Arrangement*, 95.

51. Lord Kelvin, *Baltimore Lectures on Molecular Dynamics and the Wave Theory of Light* (London: C. H. Clay and Sons, 1904), 436.

52. Diarmuid Jeffreys, *Aspirin: The Remarkable Story of a Wonder Drug* (New York: Bloomsbury, 2008).

53. G. F. Russell and J. I. Hills, "Odor Differences between Enantiomeric Isomers," *Science* 172, no. 3987 (1971): 1043–1044; T. J. Leitereg, D. G. Guadagni, Jean Harris, T. R. Mon, and R. Teranishi, "Evidence for the Difference between the Odours of the Optical Isomers (+)- and (−)-Carvone," *Nature* 230 (1971): 455–456; Lester Friedman and John G. Miller, "Odor Incongruity and Chirality," *Science* 172, no. 3987 (1971): 1044–1046.

54. Günther Ohloff, "75 Jahre Riechstoff- und Aroma-Chemie im Spiegel der Helvetica Chimica Acta," *Helvetica Chimica Acta* 75 (1992): 1341–1415 and 2041–2108; Ronald Bentley, "The Nose as a Stereochemist: Enantiomers and Odor," *Chemical Reviews* 106, no. 9 (2006): 4099–4112.

55. Chandler Burr, *The Emperor of Scent: A True Story of Perfume and Obsession* (New York: Random House, 2002).

56. Chantal Jaquet, *Philosophie de l'odorat* (Paris: PUF, 2010); A. S. Barwich, *Smellosophy: What the Nose Tells the Mind* (Cambridge, MA: Harvard University Press, 2020); Harold McGee, *Nose Dive: A Field Guide to the World's Smells* (New York: Penguin Press, 2020)

57. Elizabeth Gibney, "Force of Nature Gave Life Its Asymmetry," *Nature News*, 25 September 2014.

58. Rebecca Brazil, "The Origin of Homochirality," *Chemistry World*, 25 October 2015. G. F. Joyce, G. M. Visser, C. A. A. van Boeckel, J. H. van Boom, L. E. Orgel, and J. van Westrenen, "Chiral Selection in Poly(C)-directed Synthesis of Oligo(G)," *Nature* 310 (1984): 602.

59. Adrien D. Garcia et al., "The Astrophysical Formation of Asymmetric Molecules and the Emergence of a Chiral Bias," *Life* 9, no. 1 (2019): 29.

60. Gibney, "Force of Nature."

ACKNOWLEDGMENTS

My sense of smell broke while I was writing this book. It was the first week of March 2020, before anyone realized what this meant, and while my ability to smell never disappeared completely, it was strangely altered, with phantom aromas and distorted flavors appearing and disappearing over the next two years. The ordeal has made me more appreciative than ever of the role of olfaction, perhaps the most direct and intimate way we interact with the world. It also filled me with gratitude for my exceptional doctor, Jean Gispen, as one of the many people who made writing this book possible. The full list of this community is too long to enumerate, but I want to acknowledge in particular my excellent colleagues, who read every chapter I threw at them: Erin Drew, Michael Hoffheimer, Marc Lerner, Alex Lindgren-Gibson, Dustin Parsons, Joe Peterson, Jason Ritchie, and Tim Yenter. I want to thank Kiese Laymon, Jessie Wilkerson, and the class of A+ students who were the first to lay eyes on the project: Sarah Barch, Bethany Fitts, Chriss Fullenkamp, Page Lagarde, Olivia Morgan, and Alexis Smith. The organizers, panelists, and commentators of the conferences I've been a part of have all shaped this work for the better. Thank you José R. Bertomeu, Chris Blakely, Alice Camus, Matt Crawford, Evan Hepler-Smith, Alix Hui, Jelena Martinovic, Lucas Melvin Mueller, and Érika Wicky. I am particularly grateful to Alan Rocke and another anonymous reviewer for their care and attention. Janice Audet, Sarah Caro, and Caroline Eisenmann have been insightful and instrumental. Björn, Juneau, and Clara are the best.

Illustration Credits

INDEX

Page numbers in *italics* refer to illustrations.